LOCAL MEANINGS
GLOBAL SCHOOLING

Local meanings, global schooling

DATE DUE

			PRINTED IN U.S.A.

LOCAL MEANINGS, GLOBAL SCHOOLING
Anthropology and World Culture Theory

Edited by Kathryn M. Anderson-Levitt

LOCAL MEANINGS, GLOBAL SCHOOLING
Copyright © Kathryn M. Anderson-Levitt, 2003.
All rights reserved. No part of this book may be used or reproduced in any
manner whatsoever without written permission except in the case of brief
quotations embodied in critical articles or reviews.

First published 2003 by
PALGRAVE MACMILLAN™
175 Fifth Avenue, New York, N.Y. 10010 and
Houndmills, Basingstoke, Hampshire, England RG21 6XS.
Companies and representatives throughout the world.

PALGRAVE MACMILLAN is the global academic imprint of the Palgrave
Macmillan division of St. Martin's Press, LLC and of Palgrave Macmillan
Ltd. Macmillan® is a registered trademark in the United States, United
Kingdom and other countries. Palgrave is a registered trademark in the
European Union and other countries.

ISBN 0–4039–6162-X hardback
ISBN 0–4039–6163–8 paperback

Library of Congress Cataloging-in-Publication Data Available from the
Library of Congress

A catalogue record for this book is available from the British Library.

Design by Letra Libre, Inc.

First edition: May 2003
10 9 8 7 6 5 4 3 2 1

Printed in the United States of America.

Transferred to digital printing in 2007.

TABLE OF CONTENTS

PREFACE

This book grew out of long fascination with the European model of schooling and hence an interest in sociology's "neo-institutionalism" or world culture theory. The book also coincides with a renewed interest among anthropologists in studies of policy and hence in studies of educational reform. Many of the chapters grew out of papers presented at two symposia organized by the Committee on Transnational Education and Issues of the Council on Anthropology and Education at the 2000 and 2001 meetings of the American Anthropological Association. Other authors who shared their studies and contributed to our thinking at those sessions include Bob Herbert, Inés Dussel, Maria Paz Echeverriarza, John Napier, and Vincent Lebeta.

Very special thanks go to Elsa Statzner, who organized the first symposium and who insisted that this book happen, and to Evelyn Jacob, who made extensive, insightful comments at the second symposium. Thanks, too, to Francisco Ramirez, one of the key developers of world culture theory, who let us entice him to the anthropology meetings to serve as a discussant and who has engaged eagerly in the dialogue that this book represents. Email conversations with him and with the authors in this book, particularly Susan Jungck, helped clarify ideas presented in the introductory chapter. Finally, I am greatly indebted to the Spencer Foundation, which not only funded the research that enabled me to investigate the transnational flow of educational reforms but also gave me the opportunity to meet some of the authors participating in this volume.

A WORLD CULTURE
OF SCHOOLING?

Kathryn M. Anderson-Levitt

Is there one global culture of schooling, or many? Are school systems around the world diverging from their original European sources, or are they converging toward a single model?[1] This book opens a dialogue between two very different perspectives on schooling around the world. On the one hand, anthropologists and many scholars in comparative education emphasize national variation and, beyond that, variation from district to district and from classroom to classroom. From that point of view, the nearly 200 national school systems in the world today represent some 200 different and diverging cultures of schooling. On the other hand, sociology's "institutionalists," or world culture theorists, argue that not only has the model of modern mass education spread from a common source, but schools around the world are becoming more similar over time.[2] According to world culture theory, rather than diverging, schools are converging toward a single global model.

This question matters to anthropologists because when we look at globalization—the movement of people, money, and ideas across the entire world in unprecedented volume—we wonder whether it really means that the world is becoming more homogeneous. Are we creating a global culture (a "McWorld" for the cynical), or do people create new local cultures as rapidly as global imports hit them (Watson 1997)? Are we seeing increasing uniformity, or simply diversity organized in a new way (Hannerz 1996)? The domain of national school systems is one of the richest areas for exploring questions about globalization and, in particular, world culture theory. National school systems offer an example par excellence of an institution that has spread in the last century across the globe.

Meanwhile, for educators concerned with immediate practical problems, the question of one global culture of schooling or many has critical practical implications: Are educational reformers better advised to lobby World Bank and UNESCO policy, or to work directly with teachers in a local school? Can local educators hope to change local schools to suit local needs, or are they bound by a global model that they may or may not see?

GLOBAL SCHOOLING, LOCAL MEANINGS

Many anthropologists and comparative educationists emphasize cultural differences among national school systems. We recognize, of course, that European and North American school systems developed in parallel and that schools in the rest of the world were introduced by Europe or North America through colonial processes. However, schools introduced into new areas by different colonizers, for example, England versus France, have looked different from the beginning (Cummings 1999). Moreover, even if reformers, missionaries, or colonizers drew on some common sources when introducing schools, we have argued, schools inevitably come to reflect national culture (Spindler and Spindler 1987; Tobin, Wu, and Davidson 1989) and local people transform them to reflect local realities (for example, Flinn 1992).

In contrast, world culture theorists John Meyer, Francisco Ramirez, John Boli, and their many colleagues argue not just that the idea of schooling spread from a common source, but that schools around the world are becoming more similar over time. The world culture approach is a grand sociological theory about modern nation-states. Its theorists argue that a single global model of schooling has spread around the world as part of the diffusion of a more general cultural model of the modern nation-state, a model that also includes templates for organizing government, health systems, the military, and other institutions (Meyer et al. 1997). According to world culture theory, the global model of mass education arose in Europe as part of a state-building process (Ramirez and Boli 1987; Soysal and Strang 1989). As new nations sprang up after World War II, the rest of the world adopted the model. However, importantly, for world culture theorists it is not as if a common European model or form spread around the world once and for all. Rather, countries have re-formed their school systems over the course of the twentieth century in ways that make them more similar than they used to be. World culture theorists see "an increase in common educational principles, policies, and even practices among countries with varying national characteristics" (Chabbott and Ramirez 2000:173; see also Meyer and Ramirez 2000). Thus, for example, elementary curricula became somewhat more similar to one another from the 1920s to the 1980s (Meyer, Kamens, and Benavot 1992), and official national goals for education became somewhat more alike from 1955 to 1965 (Fiala and Lanford 1987; McNeely 1995). Rather than diverging, schools are converging toward a common model.

World culture theory makes an argument that anthropologists cannot ignore. Have human beings really created new human universals in the last 200 years? Can it be true that with a population that has grown exponentially, human beings nonetheless live in more similar societies than ever before? Alternatively, do the similarities in institutional forms and official ideologies paper over differences in everyday experience that matter a great deal more than the common framework? In practical terms, the question boils down to asking where the action is. Does true school reform happen at the level of global and national policies, or does real change happen at the level of classrooms and schools?

To address these questions within the domain of schooling, we cannot simply look for similarities or differences in schooling across countries, for whether one recognizes similarities or differences depends on the level of abstraction of the analysis. At a general level, for example, teachers often use similar repertoires of lecture, recitation, and seatwork. Yet, a more detailed analysis shows that they use the same repertoires to produce very different kinds of lessons (Anderson-Levitt 2002a). However, since world culture theory argues that school cultures are converging, whereas much of the anthropological and comparative literature implies divergence, we *can* assess the strength of each perspective by examining the direction of changes in schooling over time. Although historians of education are better positioned than anthropologists to identify long-term divergence or long-term convergence, anthropologists do study processes over periods of months or years. We can report on what has happened to ideas since they arrived from elsewhere. We can also report on whether diverse reform efforts are converging or diverging within specific sites.

With those goals in mind, this book brings together case studies—from Brazil to China and from the United States to South Africa—to scrutinize the questions raised by world culture theory. Our case studies use ethnographic and other qualitative methods to describe what is happening on the ground in particular schools and district offices and ministries of education in comparison with the reforms proposed by international agencies. We use the ethnographic details of everyday life to challenge world culture theory—showing, for example, that inside local schools, inside ministries, or even among global reformers like UNESCO and the World Bank, policy is much less homogeneous than world culture theory might imply. We show that teachers and other local actors sometimes resist and always transform the official models they are handed. We also note that world culture theorists grossly underestimate the importance of power, sometimes mistaking coercion for voluntary adoption. Nonetheless, we recognize that, by looking at the whole world at once, world culture theorists have noticed an important phenomenon that anthropologists of education miss when we focus on the local. The global view does reveal models that affect educators in local situations. Hence, this book embraces many insights from world culture theory, seeking to integrate them with what we know about lived cultures of schooling.

SCHOOLING'S COMMON FORMS

World culture theorists start from the assumption that nation-states are culturally constructed or "imagined" (Anderson 1991) rather than shaped by power struggles or economic conditions alone. Because the theorists focus on shared norms and ideas, they do not claim that the global model is necessarily the best way to run a nation—or its schools; what matters is that actors *perceive* it as the best or at least as the only acceptable way (Ramirez and Boli 1987:3). On this point, world culture theorists differ from modernists and functionalists, who tend to assume that school practices serve a society's interests and that reform usually means progress. Because world culture theorists ignore the question of power, they also differ sharply from "world systems" theorists, who argue that a global agenda is promoted through economic pressure from the World Bank, the United States Agency for International Development (USAID), Japan, and other aid donors (Arnove 1999; Ginsburg et al. 1990; Samoff 1999). For world culture theorists, school cultures converge not because nations give in to powerful donors but simply because nations voluntarily adopt what their decision makers view as the modern way to run schooling. The mechanism of change is imitation, albeit an imitation that states feel is necessary because of the pressures of interstate competition (Boli and Ramirez 1992).

What shape does the hypothesized global model of schooling take? World culture theorists point to "isomorphism" or similarity across nations in a number of elements of the schooling system (Table I.1). Some elements of the model refer to ideals stated in laws and official documents, such as the value of education as a universal right. Thus, national documents now tend to agree that education of the entire population serves the national interest (whereas in an earlier era rulers felt that it was dangerous to educate the

Table I.1 Hypothesized Common Model of Schooling

Ideals	education as a universal human right
	belief that education can have real and positive effects
	goals of education: productivity/economic growth, national development (and for a growing minority, individual development)
Basic structure	universal increase in female participation in schooling
	mass, compulsory education
	national education ministries (centralized educational policy)
	collection of educational statistics
Educational institutions	"the principle of the classroom": "egg-carton" schools with graded classes
	coeducation rather than separate schooling by ethnicity, class, gender
Content and instruction	core elementary curriculum
	predominantly whole-class lecture and recitation with seatwork

masses). Virtually all countries now subscribe to antidiscrimination policies of education (McNeely 1995) and, besides making at least token efforts to provide schooling for ethnic minorities, have opened schooling to girls (Boli and Ramirez 1992). Nations increasingly tend to agree on two official educational goals, namely, to encourage economic development and to encourage national development (as opposed to preparing students for world citizenship, training them in political ideology, offering religious training, or other goals). A growing number of nations, although still a minority, also officially claim that education should develop the individual student as well as the nation (Fiala and Lanford 1987).

Other elements of the proposed global model refer to institutional forms and practices rather than ideals. For example, every nation has instituted a mass primary schooling system in which gross enrollments are rising rapidly if they have not already reached 100 percent, and female participation is not just an ideal but a growing reality (UNDP 2000). A trend toward national laws making schooling compulsory began in the early nineteenth century for European nations and in the early to mid-twentieth century for the rest of the world (Ramirez and Ventresca 1992). Since 1910, most countries have established a central education authority—with the United States the last to join this particular bandwagon. One function of the central authority is to collect the sorts of educational statistics one can find in United Nations and World Bank reports (Ramirez and Ventresca 1992).

At the level of the local educational institution, the entire world seems to subscribe to "the principle of the classroom" (Meyer and Ramirez 2000:125). This form of organization contrasts with the one-room schoolhouse, with monitorial schools (Vincent 1980), with one-on-one apprenticeships, and with the many other ways in which peoples have organized formal education in other times and other places (Henry 1976). The principle of the classroom means that we tend to find "egg-carton" schools, in which children are clustered into a number of graded classrooms, usually with one teacher per classroom. (In small schools, however, a teacher may face more than one distinct grade in the same classroom.) There is a tendency toward age grading, although where school participation rates are relatively low, such as on the continent of Africa, one may still find an age range of several years within one grade. In general, there is a movement away from separate schooling by ethnicity and social class (Meyer and Ramirez 2000). Indeed, in this volume Diane Brook Napier describes how South Africa is moving away from one of the last school systems separated by "race." Coeducation seems to be another common if not universal feature of elementary schools. Even in countries where female seclusion is important, such as Pakistan, separate girls' schools have apparently not been common (Herz, Subbarao, Habib, and Raney 1991:29), although in a few countries there may be some movement back toward single-sex schools (Morrell 2000).

One surprising element of a proposed global model is a common core within elementary curricula around the world. World culture theorists have identified a pattern in official statements of elementary curricula, although secondary curricula are not as uniform (Kamens, Meyer, and Benavot 1996). The official core elementary curriculum in virtually every country consists of language arts, mathematics, social sciences, natural sciences, aesthetic education, and physical education; the first three subjects were already virtually universal before World War II, aesthetic and physical education became virtually universal after the war, and natural sciences became universal in the 1970s–1980s (Meyer, Kamens, and Benavot 1992). On the other hand, not all nations include religious or moral education, vocational or practical education, and hygiene, although many do. Moreover, one can predict the relative importance of the core subjects while acknowledging occasional but diminishing regional variation: about 33 percent of school time allocated to language arts, 18 percent to math, and 5 to 10 percent for each of the other core subjects (Meyer, Kamens, and Benavot 1992). The global elementary curriculum tends *not* to focus on, say, learning local geographic features, learning about food and clothing, learning about death, learning to think, or learning other content noticed by anthropologists in the formal education systems of smaller-scale societies (Henry 1976). World culture theorists also believe they see more specific trends, such as a shift from the study of history and geography to U.S.-style "social studies,"[3] a movement toward more emphasis on the world as opposed to the nation in civics education, and the incorporation of formerly elite and controversial topics in mathematics and science into mass education (Meyer and Ramirez 2000).

World culture theorists have rarely described what actually happens inside the classroom. However, Gerald LeTendre and his colleagues propose to extend world culture theory (or "institutional isomorphism," as they call it) to teaching practice and teacher beliefs (LeTendre, Baker, Akiba, Goesling, and Wiseman 2001). Drawing on their own analysis of data from the Third International Math-Science Study (TIMSS) and from the cross-national study conducted by Lorin Anderson and colleagues (Anderson 1987; Anderson, Ryan, and Shapiro 1989), they argue that teachers around the world tend most often to use whole-class lecture and recitation with student seatwork.

Supporting Evidence from Ethnographic and Comparative Studies

Anthropologists, historians, and comparative educationists know very well how schools vary on the ground. Nonetheless, if we stick to a high level of abstraction, we might agree with most of the items in Table I.1. For example, we have witnessed egg-carton schools with face-front classrooms emphasizing lecture and recitation; only in the more affluent schools in the most affluent countries do we tend to find countermodels within which children, for example, work clustered at tables or spend time in learning centers.

In fact, anthropologists might even hypothesize additional "isomorphisms," such as an official ban on corporal punishment in many if not all countries. (As in world culture theory's discussion of the official curriculum, we recognize that practice may vary dramatically from official policy.) We might also note transnational parallels in specific subject-matter pedagogy. For example, in the realm of methods for literacy instruction, teachers everywhere teach reading and writing together, not sequentially as in the past (Chartier and Hébrard 1989), and despite forceful debates between phonics and reading-for-meaning proponents, teachers tend to use a mixed method that combines both (Anderson-Levitt 2000).

COMMON REFORMS

Besides noting these commonalities in the basic forms of schooling, world culture theorists point to what seem to be the same *re*forms taking place in different countries (Table I.2). To begin with, they point out that mass schooling systems continue to expand, noting the movement toward mass secondary education in many countries and even the suggestion of movement toward mass university education (Chabbott and Ramirez 2000; Meyer and Ramirez 2000). They see long-term trends toward more nation-level control of schooling, calls for decentralization notwithstanding (Meyer and Ramirez 2000). Citing their own work in Namibia, Meyer and Ramirez see a movement toward "the nominally professionalized and somewhat autonomous teacher" (2000:126), as evidenced by the demand for higher credentials for teachers and a decline in specialized teacher training institutions. (They note a decline in other kinds of specialized vocational education as well; Meyer and Ramirez 2000.) Venturing to comment on what happens inside classrooms, they hypothesize an increasing interest in learner-centered pedagogy (McEneaney and Meyer 2000), active learning (Meyer and Ramirez 2000), and small cooperative learning groups (Ramirez 1998). World culture theorists also point to an increased concern with making the content of instruction relevant and meaningful to learners, with a resulting increase in the use of local languages in the classroom (McEneaney and Meyer 2000; Meyer and Ramirez 2000).

Case study research, including many chapters in this book, provides additional evidence that the reforms world culture theorists point to are indeed happening—or are at least being called for—in a number of different countries. We can even add to the list of hypothesized transnational reforms in Table I.2. For example, we know that mass schooling is extending its reach not only to secondary and university students, but also to children below the age of six, as Christina, Mehran, and Mir (1999) illustrate for the Middle East. However, some of our additions to the list will challenge the notion of a uniform and coherent set of reforms.

Standardization but also Decentralization

Our research confirms the claim of world culture theorists that there are movements toward increasing national control and standardization in many countries. Some countries are seeing standardized testing appear for the first time or spread to younger grades; for instance, France began nationwide testing of third and sixth graders in 1989. Assessment has come to preoc-

Table I.2 Transnational Reforms and Reform Debates

Expansion of schooling
movement toward mass secondary
movement toward mass university education
expansion of early childhood education

Decentralization	*but also*	*Standardization*
decentralization of services, site-based management		educational standards, standardized testing, quality assurance,
school choice, market, or "liberal" reforms		performance-based management, local accountability
Teacher autonomy	*but also*	*Control of teachers*
teacher professionalism and autonomy		deprofessionalization, detailed national curricula, mandated textbooks, scripted lessons
Student-centered instruction	*but also*	*Content-centered instruction*
learner-centered pedagogy, "participation," democracy in the classroom		content-based reforms, e.g., Core Knowledge, standards movement
"active learning," "hands-on" learning, projects		
small cooperative learning groups		
relevance of content to child's experience, emphasis on child's interests		
increased use of local languages		increased teaching in world languages, esp. English
reading for meaning		focus on skills in reading instruction

cupy universities in the United States and elementary schools in Namibia, as Ramirez notes (1998). In this volume, Susan Jungck points to a movement toward educational standards, standardized testing, and external quality assurance in Thailand. She also notes pressure toward performance-based management and local accountability to the Ministry of Education. Thomas Hatch and Meredith Honig in this volume describe the movement toward greater accountability to national standards in the United States. Meanwhile, Huhua Ouyang, also in this volume, points out that standardization is a long-standing practice in China.

However, in a number of countries we have also witnessed reforms that seem to contradict the trends identified by world culture theorists. Therefore, I have divided Table I.2 into two columns to represent these contradictions or tensions. To begin with, whereas world culture theorists dismiss reformers' calls for decentralization, arguing that "experiments with decentralization need to be placed in historical perspective and are not likely to result in a permanent and thorough de-nationalizing of education" (Meyer and Ramirez 2000:125), we have witnessed decentralizing reforms that seem to be having real impact in some nations. For example, while Jungck's study mentions centralization in Thailand, it also describes a movement toward decentralization and site-based management. Brook Napier comments on the devolution of authority and decentralization that has accompanied South Africa's outcomes-based education reform. There is also movement toward decentralization in France (Alexander 2000) and in Argentina (Dussell 2000).

Along the same lines, we have witnessed the neoliberal movement toward "choice" and the "marketization" of schooling, which implies decentralization, as a powerful force in many countries. In this volume, Amy Stambach describes a modest local movement toward "choice" in Tanzania, Hatch and Honig describe four schools of choice in the United States, and Lisa Rosen analyzes the movement toward the marketization of education in one U.S. community. Significantly, Lesley Bartlett, also in this volume, points out that the neoliberal pressure to offer parents' choice originates in Brazil from the World Bank, one of the international organizations to which world culture theorists attribute the convergence of world educational policy (Meyer and Ramirez 2000). Many other researchers have also demonstrated the impact of neoliberal reforms, including Bartlett and her colleagues for the United States (2002); Agnès van Zanten and Stephen Ball for France and Britain (2000; see also van Zanten 2001); Benjamin Levin for New Zealand, Canada, the United States (Minnesota), and England (2001); María Rosa Neufeld and her colleagues for Argentina (1997).

Teacher Autonomy but also Control of Teachers

Likewise, some of our studies provide evidence for the transnational movement toward teacher professionalism observed by world culture theorists. In this volume, Bayero Diallo and I document reformers promoting increased teacher autonomy in the Republic of Guinea. Jungck's study documents increased freedom of Thai teachers to shape the curriculum, and Kalanit Segal-Levit's chapter suggests the real power of teachers to effect change in Israel. Ouyang illustrates how foreign teachers import the notion of teacher autonomy into China, and Rosen shows how discourse on teacher autonomy gets used in a debate about mathematics reform. Namibia has also experimented with greater teacher autonomy (Zeichner and Dahlström 1999).

Yet, Diallo and I also document a strong countermovement in Guinea for the scripting of classroom lessons. Whereas Meyer and Ramirez argue that "attempts to de-skill teaching are replaced by standardized models of professionalized teacher training" (2000:126), we argue that attempts to de-skill teaching *coexist* with professionalized teacher training, and that the first place to witness this contradiction is inside the United States (Cochran-Smith and Fries 2001). For a view of de-skilling transnationally, see Fischman (2001). Meanwhile, Stambach and Rosen remind us that another curb on teacher autonomy is parental pressure, which decentralization reforms tend to encourage.

Student-Centered but also Content-Centered Instruction

Again, in the domain of classroom instruction, we agree with world culture theorists on the prevalence of reforms couched in the rhetoric of learner-centered pedagogy, student participation, or democracy in the classroom. In this volume, Ouyang shows how the "Communicative Method" for English-language instruction aligns with a student-centered and "discovery" approach. Diallo and I likewise note the learner-centered philosophy aligned with the push for broad teacher autonomy in Guinea, and Brook Napier mentions that the notion of learner-centered pedagogy influenced South African reforms. Elsewhere, Richard Tabulawa (1998) describes a movement for participatory, learner-centered pedagogy in Botswana, and I have described the value placed on student participation in France and the United States (Anderson-Levitt 2002b). Reformers sometimes associate student-centered teaching with work in small groups or with "hands-on" learning and small heterogeneous cooperative learning groups, as described in Namibia by Ramirez (1998). The quest to make learning relevant and meaningful for the learners is likewise present in an increased emphasis on comprehension within the mixed method of reading instruction (Anderson-Levitt 2000), and in increased use of local languages in school, as in Mali.

But again, ethnographers have also witnessed not just widespread use of content-centered or didactic methods (Baker 1997; Kumar 1990), which might be dismissed as "old-fashioned," but also "back to basics" reform movements that emphasize the transmission of a fixed curriculum rather than student inquiry.[4] Hatch and Honig describe two schools in the United States that emphasize teacher-led instruction and "the three R's"; in fact, teachers at one school note that the entire district seems to be imitating their traditional curriculum. E. D. Hirsch's Core Knowledge program had diffused to 1020 U.S. schools by May of 2001, and the Junior Great Books program to 9500 (Northwest Regional Educational Laboratory 2001). And, of course, the "standards movement" that is part of a new centralization in the United States and the National Curriculum in England also work at countercurrent to reforms in which students drive the curriculum. As for the movement toward instruction in local languages, we have also witnessed a movement toward instruction in world languages, especially English, like the one Stambach documents in Tanzania and the one Deborah Reed-Danahay describes, amazingly, in a French school in this volume (see also Cha 1992). In the United States, a movement to increase instruction in phonics skills counters the movement toward increased emphasis on comprehension (for example, Pressley 1998).

In summary, like world culture theorists, ethnographers in this volume and elsewhere have noted reforms that seem to move many countries in the same direction. However, ethnographers have also witnessed other transnational reforms that move at crosscurrents to the first set of reforms. Does the evidence point to a single global model or to something else?

ONE INCONSISTENT MODEL
OR COMPETING MODELS?

Given the inconsistencies among reforms in Table I.2, one must ask whether these reforms really do represent different parts of a single model or whether they represent different models in competition. Now, world culture theory does not demand that the global model it posits be coherent. Indeed, Meyer and his colleagues refer to "the rampant inconsistencies and conflicts within world culture" (Meyer et al. 1997:172), and Martha Finnemore points out deep tensions, notably between markets and bureaucracies as organizing principles (Finnemore 1996:341). Thus, one way to think of the apparent contradictions in Table I.2 is simply to dismiss them as reflections of the inevitable fuzziness and inconsistency in any human enterprise.

Anthropologists George and Louise Spindler offer a more satisfying way to think about contradictions within a culture. When analyzing their own diverse nation, the United States, they suggested that "we may express our commonalities as clearly in the framework of conflict as we do within the framework of cooperation" (1990:1). What people in the United States hold in common, suggest the Spindlers, is a shared way of talking about differences as well as agreements. They call this shared way of talking a *cultural dialogue,* by which they mean "culturally phrased expressions of meaning referent to pivotal concerns . . . phrased as 'value orientations' but . . . express[ing] *oppositions* as well as agreements" (1990:1, emphasis in original). For example, people in the United States talk about the value of individualism but also the value of community; they seem both to attribute success to hard work and to hold a cynical belief that success requires stepping on other people (1990). In the Spindlers' view, what makes a culture is not necessarily shared values but simply an agreement to disagree about specific opposed values. Their notion of cultural dialogue allows for conflict and contradiction within a group and even inside individual members of the group (1990:1–2).

Where our case studies provide examples of contradictions within the same national reform effort, then, we might explain it by saying that the reformers are carrying on a national cultural dialogue and hence applying a single, if inconsistent, model. Thus, in this volume Jungck points out that a single reform project in Thailand incorporates inconsistent themes. In fact, she cites Hallak (2000) to argue that decentralization necessarily invites calls for centralization.[5] In the United States, too, as Hatch and Honig show in their chapter, the government seeks simultaneously to encourage decentralization and local control while imposing tighter national standards and insisting on accountability. The messiness of these conflicts within nations or even within the very same reformers, then, does not challenge world culture theory.

Now, what are we to make of inconsistencies between nations? First, Jaekyung Lee (2001) suggests that sometimes there may be no true inconsistency at all: If and when nations start out from opposite positions, then national reforms may move in the opposite direction and yet converge toward a middle ground. Thus a U.S. system that tightens national standards and a Japanese system that deregulates schools might end up more similar than they began.

In a second scenario, if we assume a highly dynamic global model, we might explain some inconsistencies by arguing that one nation has "fallen behind" the other. For example, whole-class lecture and recitation may be the standard model of instruction in most countries (Anderson, et al. 1989; LeTendre, et al. 2001), but small-group instruction is encouraged in many countries and actually takes place pretty regularly in elementary classrooms in the United States (Ramirez 1998; Antil, Jenkins, and Wayne 1998). A country with largely whole-class instruction like France would, by this account, be less "modern" than the United States—and I've actually heard teachers in France make that claim (Anderson-Levitt 2002b). Indeed, common practices, such as whole-class lecture and recitation and use of the blackboard, were once promoted as the modern way to teach, but now new ideals have replaced them (Anderson-Levitt 2002b; Vincent

1980). Countries vary, according to this explanation, because when the global model of schooling changes, the latest version of the model, the "next wave," does not reach all parts of the globe at the same time. If we accept this notion of cultural lag, we can still accept the idea of a world culture of schooling, albeit one that keeps changing. However, the image of a dynamic global model offers no reason to expect that national school systems will become more similar over time or that reforms will converge; if the model keeps changing, we can expect that some countries will always be "behind."

It becomes even more difficult to argue for a single global model when a country supposedly in the vanguard starts to move "back" to an "older" institutional form. For example, consider the movement for greater teacher professionalism and autonomy noted above in Guinea, Namibia, Thailand, and elsewhere. What are we to make of movements "back" to scripted teaching within two important sources of teacher autonomy reform, the United States (Goodnough 2001) and England (Judge 1992), as Diallo and I discuss in this volume? The recent movement in the United States "back" to a more "traditional" emphasis on phonics offers another example. It moves counter to a trend toward increasing emphasis on reading for meaning not only in the United States from the 1980s to mid-1990s (Stahl 1999), but also in France since the 1970s (Anderson-Levitt 2002b), and in Guinea since about 1990 (Anderson-Levitt 2000). Is the United States simply making some kind of correction, as Lee (2001) might suggest, so that all of these countries will eventually converge on about the same proportion of phonics to reading for meaning? Will countries eventually converge on some balance between scripted teaching and professional autonomy? It seems more likely that we are witnessing a swing of the pendulum back and forth over the decades between phonics and reading for meaning, and between scripted and autonomous teachers.

We might interpret the global swing of the pendulum as evidence of a cultural dialogue, in the Spindlers' (1990) sense, taking place not within a nation but within the world educational community. (In the case of some of the tensions identified by Table I.2, "cultural debate" might be a better term than "cultural dialogue.") In that case, the hypothesized world culture of schooling takes the form of a transnational cultural debate rather than a consensual model. However, if countries share nothing but a cultural debate, there is no reason to expect worldwide convergence in school reform. We might as well expect a continued swinging of the pendulum resulting in countries regularly out of sync with each other.

If we kept stretching the notion of a world culture of schooling, then, we might stretch it far enough to incorporate many of the contradictions we've documented—at a price. Our broadened world culture of schooling experiences regular waves of change and incorporates cultural debates in which the pendulum swings back and forth. It offers no promise, then, of convergence toward a more coherent model.

However, some of our case studies offer evidence that does not fit the notion of a single world model no matter how broadly stretched.

ACTORS IN COMPETITION

The notion of a cultural dialogue or a cultural debate implies a conflict *within*—within a nation, within a global community, even within an individual. However, in many cases the debate takes place not within a group but rather between opposing groups of actors who are promoting competing reforms. For example, educational specialists within the World Bank do not necessarily promote the same reform ideas as UNESCO agents (Nagel and Snyder 1989). In Guinea, USAID does not give the same advice as France's international development agency (Anderson-Levitt and Alimasi 2001). In this volume, Bartlett shows how the model of education promoted by the World Bank and USAID competed in Brazil with another international model promoted by another transnational actor, the Catholic Church. Rosen gives a vivid portrait of opposing camps in the California "math wars."

Both Bartlett and Rosen see the conflicts they describe as local realizations of "enduring struggles," a concept borrowed from Dorothy Holland and Jean Lave (2001). Whereas the notion of a cultural dialogue or a cultural debate refers to *conflict within,* the idea of an enduring struggle implies *conflict between.* Moreover, whereas people engaged in cultural dialogue agree, by definition, on the terms of their conflict, people engaged in an enduring struggle may be fighting over the very definition of reality, about what the fight is (Holland and Lave 2001:22). At minimum, both Bartlett and Rosen imply that the conflicts they study represent not mere inconsistencies or debates within a global model, but rather the manifestation of conflicting models at the global level.

BUILDING THE STATE IS
NOT ALWAYS THE AGENDA

Bartlett's case also raises a question about another of world culture theory's tenets, the claim that the global model of schooling is about creating citizens of the nation-state (McEneaney and Meyer 2000; Meyer, 1997). In Bartlett's case, reforms promoted by the Catholic Church and by Paulo Freire's "Pedagogy of the Oppressed" movement hardly aim at state building. Segal-Levit's case of Soviet-inspiried grassroots reform in Israel in this volume suggests multiple agents of reform with multiple agendas, some of which contribute to state building and some of which do not. Reed-Danahay gives us a third case. Her chapter illustrates two situations in which the transnational agent stimulating reform, the European Union, acts to build Europeans or even world citizens, sidestepping national ministries in the process.[6]

GAPS BETWEEN THE MODEL (OR MODELS) AND ACTUAL BEHAVIOR

Not surprisingly, our case studies and many others point to huge gaps between a model (or models) and actual practice on the ground. One reason is that actors at various levels in importing nations sometimes resist a reform. At the level of top decision makers, Jungck in this volume shows us the Thai Ministry of Education deliberately turning away from the global model in favor of "local wisdom" (although, admittedly, Thai reformers have to adapt "local wisdom" to the exigencies of global curriculum forms and the world economy).[7] Local actors, too, may resist or give lip service only. Ouyang demonstrates the subtle and not-so-subtle ways in which students at China's vanguard institution for new methods of foreign language teaching resisted the Communicative Method practiced by foreign professors, despite the students' purported role as future champions of the new method. Diallo and I provide another case of teachers and inspectors resisting the opportunity for greater professional autonomy extended by reformers in Guinea.

Finally, as anthropologists are delighted to illustrate, even when educators accept imported models, they may transform the innovations so extensively that they create something new from them. Global imports get "creolized" (Hannerz 1987; Hannerz 1992) or "indigenized" (Robbins 2001). Brook Napier shows us how provincial governments transform imported reforms, subprovincial training programs transform them again, and local schools yet again. Ouyang illustrates the emergence of a Chinese ideal for English-language instruction that merges the best of Western and traditional methods. Stambach shows in this volume how school "choice" in a missionary-run school in Tanzania becomes, due to the U.S. educators' modifications and to reinterpretations by parents, something different from choice in the United States.

Granted, world culture theorists expect to see "loose coupling between policies and practices and practices out of sync with local realities" (Chabbott and Ramirez 2000:183). They usually mean for their theory to address only the official model, not its implementation, although they sometimes look to actual behavior as evidence, as when they cite school enrollments and female participation rates. In any case, it is legitimate to ask what exactly is spreading around the world if the same words, such as "teacher autonomy" or "choice" or "student participation" (Anderson-Levitt 2002b) or "decentralization" (Bray 1999) mean different things in different places. A model diffused in name only does not have the same significance as a model that actually affects behavior all over the world.

WORLD CULTURE THEORY AND ANTHROPOLOGY

The cases we present here make it clear that world culture theorists cannot afford to ignore what happens on the ground in particular ministries of education, provincial centers, and local classrooms. First, our examples raise the possibility that local actors find multiple, competing models out there in the larger world. Second, we are not convinced that local actors borrow models freely; hints of resistance by ministries of education suggest otherwise, and even where ministries import willingly, teachers often experience reforms as imposed from above. Third, our cases show that enacted policy differs from official policy and that this difference matters. Researchers cannot delude themselves that they are looking at the same model just because educators use a common vocabulary, while reformers cannot blithely assume that innovations will or can or should be implemented unchanged (Datnow, Hubbard, and Mehan 2002; Jacob 1999; Jennings 1996).

Do we see evidence that schooling is converging toward a single model? We report here some examples of the same kind of reform *talk* heard in very different locations: South Africa, Thailand, the United States, Guinea. However, we note that reform talk is often part of a debate rather than a homogeneous model. We also show that educators on the ground transform the meanings of the common talk or, as noted, resist it entirely. Moreover, we have provided examples of alternative sources of ideas (the left-leaning wing of the Catholic Church, the European Union, Soviet immigrants) diffusing alternative kinds of reforms. From our vantage points, we do not see a world becoming more homogeneous with time.

Nonetheless, anthropologists must recognize the importance of insights from world culture theory. First, the theory offers a plausible explanation of the very phenomenon anthropologists in this volume are studying, the presence of modern schools around the world. Second, although there may be several rather than one transnational model, world culture theory has nonetheless identified a striking set of isomorphisms. None of us can ignore that ministries of education, school inspectors, teachers, students, and parents import, play with, or react against a set of similar-looking reforms that are traded back and forth across countries. Third, though elements of transnational forms and reforms mean different things in different local settings, they are far from meaningless (Anderson-Levitt 2002a). The common educational discourse and taken-for-granted features of school structure tend to set limits on our ordinary thinking about schooling. Researchers who focus only on the local or who see only cross-national differences are missing the iceberg under the surface. Reformers who don't see the global model or models beneath local differences are tinkering only with the surface.

If we take seriously both local variability and world culture theory, we recognize that each perspective on its own misses something crucial. It follows that we cannot simply shift the focus from the transnational to the local as we please. We must view schooling from both perspectives simultaneously, as Stambach argues in this volume.

A complete theory of schooling and of school reform would begin by acknowledging that there is a common set of models of modern schooling— or a common set of cultural debates about schooling. Contemporary mass schooling is an institution with (not completely homogeneous) roots in Europe. It spread to the rest of the world, as world culture theorists argue, as part of the enterprise of constructing modern-looking nation-states in the transnational system. At a very general level, schooling manifests a set of common forms, as identified in Table I.1 above. Although there seem to be some competing transnational models, such as Freire's, very few really qualify as competition at this level of the general form.

At the same time, as many of the case studies in this volume show, administrators, teachers, and students create within the roughly common structure very different lived experiences (Anderson-Levitt 2002a). This happens in part because there are different actors at different levels with different agendas. In addition, there are inevitably gaps between model and practice. The enacted curriculum in any school in the world can contrast sharply with the official curriculum. Teachers or even entire nations play with the common classroom repertoire, to "indigenize" the structure or tone of classroom life in ways that make it either a caricature of schooling (Fuller 1991; Watson-Gegeo and Gegeo 1992) or a better place for local children (Erickson and Mohatt 1982; Flinn 1992).

In many ways, differences in lived experiences matter more than the common structure. Nevertheless, ultimately, the fact of the common structure matters. It puts a frame around ordinarily thinkable ways of doing schooling. By implication, there are three kinds of reform movements. Some reforms, such as the extension of schooling to all children or the press for efficiency of execution, spread or strengthen the common models. A second kind of reform works within the common models to change the lived experience of students or of teachers, as in Segal-Levit's example of new kinds of science teaching or other successful efforts at such activities as "active learning" or teacher autonomy. Finally, there are rare reforms that would remove us from the contemporary common model, such as Freire's vision or Ivan Illich's proposal to de-school society. The last kind of reform is truly radical and very difficult to bring about, although changes of this magnitude have occurred over the centuries. It behooves reformers to know which kind of reform they are attempting.

ORGANIZATION OF THIS BOOK

Most of the case studies in this book look at elementary schools, but Rosen and Segal-Levit focus on secondary schools, Ouyang looks at higher education, and Bartlett at adult education. Most raise serious questions about world culture theory; Bartlett in particular provides an extensive critique. At the same time, several of the studies note the value of world culture theory and the significance of transnational models—notably Brook Napier, Anderson-Levitt and Diallo, and Rosen. Hatch and Honig's study indirectly supports world culture theory by illustrating the pressures for isomorphism. Several authors take us beyond world culture theory to alternative theories that cast other lights on global school reform: Jungck uses Roland Robertson's notion of "glocalization," Stambach discusses Andrew Strathern's view of the interplay of local and universal narratives, Rosen and Bartlett bring in Holland and Lave's "enduring struggles," Bartlett develops a concept of educational projects derived from Omi and Winant, and Reed-Danahay demonstrates the usefulness of Foucault's notion of discipline in the context of transnational school reform.

In Part I of this volume, three chapters trace the movement of reforms from ministries of education to local schools. Jungck examines one part of Thailand's National Education Act of 1999, which prescribed the incorporation of "local wisdom" into schools to counter the heavy presence of a global model of schooling. Her examples illustrate the combination of global and local forces in particular schools as teachers attempt to use local wisdom. Brook Napier traces the movement of South Africa's complex set of reforms from the national ministry to provinces, subprovinces, and local schools, noting international influences at every level. Anderson-Levitt and Diallo consider the movement for greater teacher autonomy in the Republic of Guinea promoted by U.S. reformers and by some World Bank reformers. We show that teacher autonomy is an idea contested in source countries, transformed by importing officials, resisted by local teachers, and yet practiced, sometimes unwittingly, in the classroom.

In Part II, four chapters further explore local levels, examining reforms as experienced by teachers, students, and parents. Hatch and Honig show that even in a decentralized nation committed to educational alternatives—the United States—truly distinctive schools are extremely difficult to maintain. They make us wonder whether real decentralization is even possible. Ouyang examines a different kind of "democratizing" reform, the attempt to build the Communicative Method and hence more student participation into English-language instruction in China. He explores how and why students and even faculty colleagues at Guandong's most progressive institute for the training of language teachers resisted the efforts of foreign professors whom they had specifically invited to introduce the new method. His study also illustrates how transnational models arrive loaded down with cultural baggage from the source country. Stambach explores Tanzanian parents' reactions to an English-language program that was introduced into primary schools by U.S. missionaries in 1999. In illustrating how a single program can be interpreted in different ways, it advances a framework that accounts for the simultaneous development of cultural variability and institutional homogeneity in the arena of mass education. Rosen examines the local meanings of "choice" in a very different setting, a city in the United States. She shows how one group of parents appropriated the transnational vocabulary of neoliberalism to contest teachers' autonomy to adopt a new mathematics curriculum. Significantly, Rosen also argues that local conflicts like the one she describes eventually shape larger, perhaps even transnational, struggles.

In Part III, three chapters raise questions about world culture theory by analyzing reforms that come from outside the transnational system. Bartlett contrasts models of knowledge production in contemporary Brazilian adult literacy programs. The efficiency model of the World Bank and other large donors presents education as investment in future workers, positioning education in the service of the economy, whereas the popular education common in nongovernmental-organization literacy programs embraces knowledge as power. Reed-Danahay illustrates two cases in which the transnational agent stimulating reform, the European Union, acts to build Europeans or even world citizens, not citizens of nation-states. Because the actors (from government officials to teachers to students) respond through the lenses of both regional and national identities, this transnational policy initiative does not have global, uniform effects. Segal-Levit shows how the culture of scientific education that existed in the former USSR has diffused to schools in Israel, first through the informal efforts of Soviet-born immigrant teachers and only later through official incorporation into the national school system. The effort has led to mutual fertilization and the creation of a hybrid based on the unique experience of the immigrants but adapted to the Israeli system.

Francisco Ramirez closes the book with a comment from the perspective of world culture theory, pointing out similarities between its global argument and the anthropological perspective.

NOTES

1. When I refer to "school systems," "schooling," or "schools" in this chapter, I mean government-run or government-affiliated systems intended for the entire population. This includes, for example, the Catholic school systems in France and in the United States, which function as complements to the government-run public school system, but does not include such formal schooling as Quranic schools, which operate independently from the state in regions like West Africa, or the former system of Mandarin education in China, which was reserved for the few rather than the masses.

2. The label "world culture" comes from Ramirez, Soysal, and Shanahan (1997) and from the reference to a "rationalized world institutional and cultural order" in Meyer et al. (1997:151). Other authors refer to the ideas of Meyer and his colleagues variously as "institutional theory" (Berkovitch 1999:7), "institutionalism" (Finnemore 1996; LeTendre et al. 2001), "neo-institutionalism" (Levin 2001), "global rationalization" (Davies and Guppy 1997), and, somewhat misleadingly, "world systems theory" (Cummings 1999).

3. However, France moved back to history and geography as separate subjects in its elementary curriculum in 1985.

4. As Robin Alexander demonstrates with detailed cases from five countries, the dichotomy between student-centered and content-centered, like all the dichotomies in Table I.2, is grossly oversimplified (2000, 2002). Alexander identifies six rather than two "versions of teaching" used in various combinations in these countries. See also the eight-sided layout of teaching philosophies in Peretti 1993.

5. Curiously, the progressive movement of the 1910s and 1920s in the United States encompassed a similar tension between a move toward democratic,

student-centered pedagogy and a move toward greater efficiency and bureaucratic control (Semel and Sadovnik 1999).

6. McEneaney and Meyer now suggest, on the basis of new data, that "civics instruction dramatically shifted in the 1980s and 1990s to a model of the 'postnational citizen' of the world" (2000:200), but it is not clear how this apparent change in the global model fits their larger theory.

7. McEneaney and Meyer (2000) try to explain the incorporation of local cultures and local languages within world culture theory as part of an effort to construct individual citizens engaged with a relevant curriculum. However, the picture they paint of increasingly diversified curricula does not sit easily with their claim that curricula continue to become more similar around the world over time.

REFERENCES

Alexander, R. 2000. *Culture and pedagogy: International comparisons in primary education*. Oxford, UK: Blackwell.

Alexander, R. 2002. Still no pedagogy? Paper presented at the annual meeting of the American Educational Research Association, New Orleans, April 3.

Anderson, B. 1991. *Imagined communities*. London: Verso.

Anderson, L. W. 1987. The classroom environment study: Teaching for learning. *Comparative Education Review* 31(1): 69–87.

Anderson, L. W., D. Ryan, and B. Shapiro. 1989. *The IEA classroom environment study*. New York: Pergamon.

Anderson-Levitt, K. M. 2000. What counts as the mixed method of reading instruction in Guinea? Fractures in the global culture of modern schooling. Paper presented at the annual meeting of the American Educational Research Association, New Orleans, April 24.

Anderson-Levitt, K. M. 2002a. Teaching culture as national and transnational: A response to "Teachers' Work." *Educational Researcher* 31(3): 19–21.

Anderson-Levitt, K. M. 2002b. *Teaching cultures: Cultural knowledge for teaching first grade in France and the United States*. Cresskill, NJ: Hampton Press.

Anderson-Levitt, K. M., and N.-I. Alimasi. 2001. Are pedagogical ideals embraced or imposed? The case of reading instruction in the Republic of Guinea. In *Policy as practice: Toward a comparative sociocultural analysis of educational policy*, edited by M. Sutton and B. Levinson. Norwood, NJ: Ablex.

Antil, L. R., J. R. Jenkins, and S. K. Wayne. 1998. Cooperative learning: Prevalence, conceptualizations, and the relation between research and practice. *American Educational Research Journal* 35(3): 419–54.

Arnove, R. 1999. Reframing comparative education: The dialectic of the global and the local. In *Comparative education*, edited by R. Arnove and C. A. Torres. Lanham, MD: Rowman and Littlefield.

Baker, V. 1997. Does formalism spell failure? Values and pedagogies in cross-cultural perspective. In *Education and cultural process: Anthropological approaches,* edited by G. D. Spindler. Prospect Heights, IL: Waveland.

Barlett, L., M. Frederick, T. Gulbrandsen, and E. Murillo. 2002. The marketization of education: Public schools for private ends. *Anthropology and Education Quarterly* 33 (1): 5–29.

Berkovitch, N. 1999. *From motherhood to citizenship: Women's rights and international organizations.* Baltimore: Johns Hopkins.

Boli, J., and F. O. Ramirez. 1992. Compulsory schooling in the Western cultural context. In *Emergent issues in education: Comparative perspectives,* edited by R. F. Arnove, P. G. Altbach, and G. P. Kelly. Albany: State University of New York.

Bray M. 1999. Control of education: Issues and tensions in centralization and decentralization. In *Comparative education,* edited by R. F. Arnove and C. A. Torres. Lanham, MD: Rowman & Littlefield.

Cha Y. 1992. Language instruction in national curricula, 1850–1986: The effect of the global system. In *School knowledge for the masses: World models and national primary curricular categories in the twentieth century,* edited by J. W. Meyer, D. Kamens, and A. Benavot. Washington, D.C.: Falmer.

Chabbott, C., and F. O. Ramirez. 2000. Development and education. In *Handbook of the Sociology of Education,* edited by M. T. Hallinan. New York: Kluwer Academic/Plenum Publishers.

Chartier, A.-M., and J. Hébrard. 1989. *Discours sur la lecture, 1880–2000.* Paris: Bibliothèque publique d'information–Centre Pompidou.

Christina, R., G. Mehran, and S. Mir. 1999. Education in the Middle East: Challenges and opportunities. In *Comparative education,* edited by R. F. Arnove and C. A. Torres. Lanham, MD: Rowman & Littlefield.

Cochran-Smith, M., and M. K. Fries. 2001. Sticks, stones, and ideology: The discourse of reform in teacher education. *Educational Researcher* 30 (8): 3–15.

Cummings, W. K. 1999. The institutions of education: Compare! Compare! Compare! *Comparative Education Review* 43 (4): 413–437.

Datnow, A., L. Hubbard, and H. Mehan. 2002. *Extending educational reform: From one school to many.* New York: Routledge Falmer.

Davies, S., and N. Guppy. 1997. Globalization and educational reforms in Anglo-American Democracies. *Comparative Education Review* 41 (4): 435–459.

Dussel, I. 2000. Argentina: Education reform and the construction of a national imaginary in Argentina (1880–1990). Paper presented at the American Educational Research Organization, New Orleans, April 27.

Erickson, F., and G. Mohatt. 1982. Cultural organization of participation structures in two classrooms of Indian students. In *Doing the ethnography of schooling,* edited by G. D. Spindler. Prospect Heights, IL: Waveland.

Fiala, R., and A. G. Lanford. 1987. Educational ideology and the world educational revolution, 1950–1970. *Comparative Education Review* 31 (3): 315–332.

Finnemore, M. 1996. Norms, culture, and world politics: Insights from sociology's institutionalism. *International Organization* 50 (2): 325–347.

Fischman, G. 2001. Teachers, globalization, and hope: Beyond the narrative of redemption. Essay review. *Comparative Education Review* 45 (3): 412–418.

Flinn, J. 1992. Transmitting traditional values in new schools: Elementary education of Pulap Atoll. *Anthropology and Education Quarterly* 23 (1): 44–59.

Fuller, B. 1991. *Growing up modern: The Western state builds Third-World schools.* New York: Routledge.

Ginsburg, M. B., S. Cooper, R. Raghu, and H. Zegarra. 1990. National and world-systems explanations of educational reform. *Comparative Education Review* 34 (4): 464–499.

Goodnough, A. 2001. Teaching by the book, no asides allowed. *New York Times,* 23 May, A1, 27.

Hallak, J. 2000. Globalisation and its impact on education. In *Globalisation, educational transformation and societies in transition,* edited by T. Mebrahtu, M. Crossley, and D. Johnson. Oxford, UK: Symposium Books.

Hannerz, U. 1987. The world in creolization. *Africa* 57: 546–559.

Hannerz, U. 1992. *Cultural complexity: Studies in the social organization of meaning.* New York: Columbia.

Hannerz, U. 1996. *Transnational connections.* London: Routledge.

Henry, J. 1976. A cross-cultural outline of education. In *Educational patterns and cultural configurations: The anthropology of education,* edited by J. I. Roberts and S. K. Akinsanya. New York: David McKay.

Herz, B., K. Subbarao, M. Habib, and L. Raney. 1991. *Letting girls learn: Promising approaches in primary and secondary education.* Washington, D.C.: The World Bank.

Holland, D., and J. Lave. 2001. *History in person: Enduring struggles, contentious practice, intimate identities.* Santa Fe, NM: School of American Research.

Jacob, E. 1999. *Cooperative learning in context: An educational innovation in everyday classrooms.* Albany: State University of New York Press.

Jennings, N. E. 1996. *Interpreting policy in real classrooms: Case studies of state reform and teacher practice.* New York: Teachers College Press.

Judge, H. 1992. A cross-national study of teachers. In *Emergent issues in education: comparative perspectives,* edited by R. F. Arnove, P. G. Altbach, and G. P. Kelly. Albany: State University of New York Press.

Kamens, D. H., J. W. Meyer, and A. Benavot. 1996. Worldwide patterns in academic secondary education curricula. *Comparative Education Review* 40 (2):116–138.

Kumar, K. 1990. The meek dictator: The Indian teacher in historical perspective. In *Handbook of Educational Ideas and Practices,* edited by N. Entwistle. New York: Routledge.

Lee, J. 2001. School reform initiatives as balancing acts: Policy variation and educational convergence among Japan, Korea, England and the United States. *Educa-*

tional Policy Analysis Archives 9 (13). Available at http://epaa.asu.edu. [Accessed October 31, 2001.]

LeTendre, G., D. P. Baker, M. Akiba, B. Goesling, and A. Wiseman. 2001. Teachers' work: Institutional isomorphism and cultural variation in the U.S., Germany, and Japan. *Educational Researcher* 30 (6): 3–15.

Levin B. 2001. *Reforming education: From origins to outcomes.* New York: Falmer.

McEneaney, E. H., and J. W. Meyer. 2000. The content of the curriculum: An institutionalist perspective. In *Handbook of the sociology of education,* edited by M. T. Hallinan. New York: Kluwer Academic/Plenum Publishers.

McNeely, C. L. 1995. *Constructing the nation-state: International organization and prescriptive action.* Westport, CT: Greenwood.

Meyer, J. W., J. Boli, G. M. Thomas, and F. O. Ramirez. 1997. World-society and the nation-state. *American Journal of Sociology* 103 (1): 144–181.

Meyer, J W., D. Kamens, and A. Benavot. 1992. *School knowledge for the masses: World models and national primary curricular categories in the twentieth century.* Washington, D.C.: Falmer.

Meyer, J. W., and F. O. Ramirez. 2000. The world institutionalization of education. In *Discourse formation in comparative education,* edited by J. Schriewer. New York: Peter Lang.

Morrell, R. 2000. Considering the case for single-sex schools for girls in South Africa. *McGill Journal of Education* 35 (3): 221–44.

Nagel, J., and C. W. Snyder, Jr. 1989. International funding of educational development: External agendas and internal adaptations—the case of Liberia. *Comparative Education Review* 33 (1): 3–20.

Neufeld, M. R., S. Carro, A. Padawer, S. Pallma, and S. I. Thisted. 1997. *Las escuelas del conurbano bonaerense: idas y venidas de una decada. Informe Final.* Unpublished MS. Buenos Aires: Faculty of Philosophy and Letters,

Northwest Regional Educational Laboratory. 2001.*The Catalog of School Reform Models.* Available at: http://www.nwrel.org/scpd/catalog/index.shtml. [Accessed June 12, 2002.]

Peretti, A. de. 1993. *Controverses en éducation.* Paris: Hachette.

Pressley, M. 1998. *Reading instruction that works: The case for balanced teaching.* New York: Guilford.

Ramirez, F. O. 1998. SIMs, CAMs, and gender equity: Constructing the progressive learner in the North. In *Inside reform: Policy and programming considerations in Namibia's basic education reform,* edited by C. W. Snyder, Jr., and F. G. G. Voigts. Windhoek, Namibia: Gamsberg Macmillan.

Ramirez, F. O., and J. Boli. 1987. The political construction of mass schooling: European origins and worldwide institutionalization. *Sociology of Education* 60 (1): 2–17.

Ramirez, F. O., Y. Soysal, and S. Shanahan. 1997. The changing logic of political citizenship: Cross-national acquisition of women's suffrage rights, 1890 to 1990. *American Sociological Review* 62: 735–745.

Ramirez, F. O., and M. J. Ventresca. 1992. Building the institution of mass schooling: Isomorphism in the modern world. In *The political construction of education: The state, school expansion, and economic change*, edited by B. Fuller and R. Rubinson. New York: Praeger.

Robbins, J. 2001. God is nothing but talk: Modernity, language, and prayer in a Papua New Guinea society. *American Anthropologist* 103 (4): 901–912.

Samoff, J. 1999. Institutionalizing international influence In *Comparative education*, edited by R. Arnove and C. A. Torres. Lanham, MD: Rowman and Littlefield.

Semel, S. F., and A. R. Sadovnik. 1999. *Schools of tomorrow, schools of today: What happened to progressive education?* New York: Peter Lang.

Soysal, Y. N., and D. Strang. 1989. Construction of the first mass education systems in nineteenth-century Europe. *Sociology of Education* 62: 277–288.

Spindler, G. D., and L. Spindler. 1987. Schönhausen revisited and the rediscovery of culture. In *Interpretive ethnography of education*, edited by G. D. Spindler and L. Spindler. Hillsdale, NJ: Lawrence Erlbaum.

Spindler, G. D., and L. Spindler. 1990. *The American cultural dialogue and its transmission*. Philadelphia: Falmer Press.

Stahl, S. A. 1999. Why innovations come and go (and mostly go): The case of whole language. *Educational Researcher* 28 (8): 13–22.

Tabulawa, R. 1998. Teachers' perspectives on classroom practice in Botswana: Implications for pedagogical change. *International Journal of Qualitative Studies in Education* 11 (2): 249–271.

Tobin, J. J., D. Wu, and D. Davidson. 1989. *Preschools in three cultures*. New Haven: Yale University Press.

UNDP. United Nations Development Programme. 2000. *Human development report 2000*. New York: Oxford University Press for the UNDP.

van Zanten, A. 2001. *L'école de la périphérie*. Paris: Presses Universitaires de France.

van Zanten, A., and S. Ball. 2000. Comparer pour comprendre: Globalisation, réinterprétations nationales et recontextualisations locales des politiques éducatives néolibérales. *Revue de l'Institut de Sociologie* 1 (4): 113–131.

Vincent, G. 1980. *L'Ecole primaire française: Etude sociologique*. Lyon: Presses Universitaires de Lyon.

Watson, J. L. 1997. *Golden arches east: McDonald's in East Asia*. Stanford, CA: Stanford University Press.

Watson-Gegeo, K. A., and D. W. Gegeo. 1992. Schooling, knowledge, and power: Social transformation in the Solomon Islands. *Anthropology and Education Quarterly* 23 (1): 10–29.

Zeichner, K., and L. Dahlström. 1999. *Democratic teacher education in Namibia*. Boulder, CO: Westview.

CHAPTER ONE

"THAI WISDOM" AND GLOCALIZATION

Negotiating the Global and the Local in Thailand's National Education Reform

Susan Jungck
with Boonreang Kajornsin

About a hundred years ago, King Rama V strategically thought of ceding power over some small areas of land to Britain and France to protect the greater part of the Kingdom, instead of fighting in the battlefield where traditional weapons could not compete with the more advanced war equipment of the Europeans. When reinforced by implementing a strategic alliance with Russia to balance the superpowers, the country was saved and became the only one in the region never to have experienced being colonized. However, in the dynamic pattern of the economic battleground in the era of Globalization when "information and currencies move across the national border at lightning speed" (Prawase 1998), the country was defeated. Historic Thai wisdom, containing the ability to think strategically, had been lost. With weakness of the macro-economic structure . . . the country collapsed on July 2, 1997, and the negative results of this collapse were felt by all Thai. (Supradith Na Ayudhya 2000:1)

In this paper, I argue that the "Thai wisdom" supposedly lost is emerging as a national phenomenon, a significant form of response to the forces of globalization and a perceived "world culture." Revitalizing Thai wisdom has become part of the comprehensive educational reforms that followed Thailand's economic crash of 1997—the National Education Act of 1999. Recourse to Thai wisdom, and more specifically *local wisdom,* is a strategy aimed at revitalizing cultural knowledge and practices. I believe that this revitalization represents a form of resistance to the homogenizing aspects of globalization enacted through movements to enhance the capacities of local communities to chart their own cultural and economic paths.

The significance of the current educational reform projects that I will reference is that they are stimulating community participation processes for revitalizing Thai wisdom and developing locally relevant curricula. Furthermore, as the government begins to stall on implementing the new reforms, the fate of educational change in Thailand will increasingly reside with civil society and community-based projects like these, which are explicitly intended to promote local voice. Education projects like the ones I will reference are critical sites relative to globalization because they occasion *glocalization,* where tradition and modernity are (re)invented (Lukens-Bull 2001) through the interplay of local and global forces.

Rung Kaewdang, the secretary-general of Thailand's Office of the National Education Commission (ONEC), the office responsible for educational policy development, is a major architect of the new reforms. Conceptualizing the revitalization of Thai wisdom as a cultural strategy, he notes:

> In the past forty years . . . Thailand's economic and social development . . . depended too much on Western knowledge and know-how. . . . The economic crisis that has occurred during the past three years was the outcome of such mistakes and caused us to reconsider, review and re-evaluate our social and economic development plan. . . . We discovered that we had pursued Western ways of development and entirely neglected our own indigenous or local knowledge, the splendid treasure that has played important roles in building the nation's unity and dignity. Now it is the time we should turn back to our own philosophy, our own culture, and our own indigenous knowledge. . . .
>
> Education in the globalization age should therefore be the balanced integration between global knowledge and indigenous knowledge. (Kaewdang 2001a)

GLOBAL REFORMS, THAI RESPONSES

Many argue that education systems worldwide are embracing rational global reform paradigms, creating "considerable isomorphism in schooling" characterized by "marketization, individualization, bureaucratization, and homogenization" (LeTendre et al. 2001). Indeed, many Asia-Pacific countries, such as Thailand, South Korea, Japan, Hong Kong, Malaysia, Indonesia, Singapore, New Zealand, and Australia, have adopted education reforms with similar themes: decentralized systems of administration and accountability; more community-level participation; teacher reform; higher education reform; new technologies; private sector involvement; educational standards, quality assurance, and performance-based assessments; and more learner-centered curricula (Cheng 2001; Hallak 2000; Kwak 2001; Lee 2001; ONEC 2001; Pascoe 2001). In part, these reforms reflect the influence of global capitalism, through which liberalization, reduced government spending, privatization, decentralized bureaucracies, and corporate managerialism are driving policies. For example, after the economic crash in 1997, international aid agencies mandated that Thailand focus on improving productivity, internal and external competition, the country's knowledge base, capability in science and technology, and political management in line with the tenets of "good governance" (*The Nation* 2001b; Supradith Na Ayudhya 2000). Like other countries in debt, Thailand had little option but to adopt these policies (Madrick 2001; Bowornwathana 2000). However, that does not mean that the adopted policies are universally, genuinely, or unambiguously embraced.

Bidhya Bowornwathana (2000) argues that in Thailand, reform drafters and politicians have insufficient time and expertise to develop indigenous solutions to major social, economic, and educational problems. Therefore, he claims:

> Adopting a global reform paradigm is a good choice, from this perspective, because it silences domestic differences, pleases funding agencies, and presents convenient packages of ready-made reform programs. It is also easier to convince the public about the benefits of a reform proposal that has already worked well in a developed country than to build public support for a completely new indigenous reform program. (Bowornwathana 2000:398)

Furthermore, these reformers falsely assume that a successful reform strategy in one country can be transferred to other countries regardless of their contextual differences. Consequently, Bowornwathana argues that these national reform policies are unsuccessful in the long run. Even though governance reforms are still at an early stage in Thailand, he claims that "Thai citizens are not especially benefiting from the public reform initiatives" (Bowornwathana 2000:394). Though Bowornwathana was speaking specifically of Thailand's national governance reforms, the new education reform policies reflect the same mandates and dilemmas, and foreshadow the same fate. The focus of this paper is on some local reform strategies, which, in spite of the above context, are nonetheless working to ensure that education will be contextualized in Thai wisdom.

THE CONTEXT OF EDUCATIONAL REFORM IN THAILAND: THE ECONOMIC CRASH AND A NEW CONSTITUTION

Critics had long claimed that Thailand's highly centralized, top-down, "chalk and talk" education system, with its standardized curriculum, rote learning, and high-stakes testing, had to go. Students, they said, needed to develop at their own pace and teachers needed to facilitate analytical, creative, and independent thinking skills. However, reform became a national priority only after the economy crashed in 1997 and the new Constitution mandated that the system be changed. The immediate causes of the economic crash were attributed to a lack of governmental financial discipline, adverse development, and a series of currency attacks (*The Nation* 2001a; Supradith Na Ayudhya 2000; *The Economist* 1997; IMF 1997; Neher n.d.). However, social critics diverge on the long-term causes and implications of the crisis, as Redmond describes:

> The Asian economic collapse that began in Thailand in July 1997 has promoted two prominent responses—utterly different, and utterly predictable. One is that "the West knows best," that Asia has stumbled because it failed to absorb enough of Euro-American capitalism, democracy, and the individualistic ideas that support them. Others insist that the very suddenness of the reversal points to the opposite conclusion, that the wholesale consumption of foreign funds and value systems led to greed, blindness, and a breakdown in the moral balance and social control Asian societies thrive on. (Redmond 1999: back cover)

There has been much more consensus however, that one of the long-term causes of the crisis was an inadequate education system (Zack 1997; *The Nation* 2001a). Hence, educational reform has become a mandated remedy:

> As educators, we cannot deny the responsibility for the economic, social, cultural and political ill effects since people who caused all these problems are the products of our current educational system. (Kaewdang 2001b)

A new Thai Constitution aimed at extending civil liberties through decentralizing power and restructuring controls over the governmental, political, and educational systems was approved shortly after the economic crash in 1997 (Phongpaichit and Baker 1998). The Constitution calls for more participatory democracy and contains an unprecedented number of provisions related to education, notably that the educational system and the teaching profession will be improved consistent with economic and social changes and that every citizen has the right to a 12-year basic and free education. Finally, the Constitution mandated that a new national education law be developed within two years.

THE NATIONAL EDUCATION ACT OF 1999

Two years after the economic crash and the adoption of the new Constitution, the National Education Act of 1999 was passed. Implementation was to be phased in and completed by 2002. This comprehensive Act addresses: teaching and learning; the revitalization of Thai wisdom; the empowerment of teachers; more student-centered instruction; administrative and fiscal decentralization; and a system of educational standards, quality assurance, and authentic assessment (Kaewdang 2001b). Encoded in the Act are two divergent themes (Jungck 2001; Cheng 2001). On the one hand, the Act calls for a tightly coupled system with more effectiveness, economic efficiency, accountability, measurable outputs, quality assurance, and standardization:

> There shall be a system of educational quality assurance to ensure improvement of educational quality and standards at all levels. Such a system shall be comprised of both internal and external quality assurance. (Section 47, ONEC 1999)

On the other hand, the Act envisions a loosely coupled system with more decentralized authority and decision making, local participation, curricular relevance, and student-centered learning:

> The Ministry shall decentralize powers in educational administration and management regarding academic matters, budget, personnel and general affairs administration directly to the Committees and Offices for Education, Religion, and Culture of the educational service areas and the education institutions in the (local) areas. (Section 39, ONEC 1999)

According to Hallak (2000:25), it is not so unusual that countries embrace seemingly contradictory policies like the above, as one of the consequences of globalization is that it generates both standardizing and diversifying phenomena. I see both standardizing and diversifying phenomena thematically reflected in Thailand's new educational reform policy: One theme is about democratizing and devolving power and the other is about developing ways to manage and assess efficiency and effectiveness. Watson (2000) suggests that when highly centralized systems move to less rigid forms of control, issues of how to assure equity and quality typically surface. As one consequence, states introduce centralized educational standards and monitoring systems (Watson 2000). Thus, decentralization seems to inevitably trigger, as it has in Thailand, more state standards and monitoring. The central dilemma is that reforms aimed at democratizing and decentralizing highly centralized government systems require long-term, developmental, constructivist, and high-risk innovations involving local capacity building, curricular innovation, transformative leadership and greater school diversity, while reforms aimed at standardization and performance-based managerial systems tend to thwart the formative nature of these long-term processes and changes.

Many believe that the Act's emphasis on administrative and academic decentralization is especially vital for rural people, who constitute a 60 percent majority in Thailand, because the traditional curriculum has been largely irrelevant for them (Thongthew 1999; Kaewdang 2001a). The Act specifically emphasizes the importance of learning in the communities themselves through engendering more community participation in school affairs. While decentralization is intended, in part, to increase efficiency and local accountability, many reformers advocate it as a means of giving local stakeholders more voice and influence in their schools. Stakeholders, they say, can "benefit from local wisdom and other sources of learning for community development in keeping with their requirements and needs" (Section 29, ONEC 1999).

As Rung Kaewdang envisions, "As schools will have more autonomy to decide the local curriculum they deem necessary for local children, there is a possibility that Thai wisdom will enjoy the same status as modern knowledge. Our children and adults will learn to be Thais in parallel with the internationalization" (Kaewdang 2001b). Consequently, he explained, ONEC has researched "Thai knowledge" in order to "revitalize and return it to our educational system":

> [L]ocal wisdom enables lifelong learning in society. It not only strengthens the community's economic situation on the basis of self-sufficiency, but also moral values, and local culture among community people. In the globalized world, it is certain that most of the contents in the Internet will focus on the Western knowledge, ideas, and culture. However, if there is nothing done to promote the learning of local knowledge, our future generations

will definitely not understand where we are in the world or even lose the root of their culture. Education in the globalization age should therefore be the balanced integration between global knowledge and indigenous knowledge . . . for sustainable development in any community, international understanding, and peace and harmony of the world. (Kaewdang 2001a)

In this paper, I suggest that revitalizing Thai wisdom has become a national strategy to encourage a more balanced mediation of "global knowledge and indigenous knowledge."

THAI WISDOM/LOCAL WISDOM

Rung Kaewdang defines "Thai wisdom" and "local wisdom" as "the bodies of knowledge, abilities, and skills of Thai people accumulated through many years of experience, learning, development, and transmission. It has helped to solve problems and contribute to the development of our people's way of life in accordance with the changing times and environment" (Kaewdang 2001a).

Thai wisdom has become a national phenomenon: It is a popular cultural campaign, sanctioned by law and firmly embedded in the new educational reforms. While the term or concept of Thai wisdom is not new, referencing and encoding it in national policies is (Srisalab and Sritongsuk 1999). In 1997, the right and the duty to promote Thai wisdom was incorporated in the new Constitution:

> Section 46: Persons so assembling as to be a traditional community shall have the right to conserve or restore their customs, local knowledge, arts or good culture of their community and of the nation. . . .
> Section 81: The state shall promote local knowledge and national arts and culture.
> . . . .
> Section 289: A local government organization has the duty to conserve local arts, custom, knowledge and good culture. (Kaewdang 2001a)

Thai wisdom has become an umbrella theme linking sociocultural resources with economic growth and development. For example, the stated intent of the Office of the National Cultural Commission's project "Thai Culture Fights off Economic Crisis" was to "help strengthen individuals and communities in their determination to become self-reliant economically, psychologically and socially and to rediscover the values of Thai traditional wisdom with a sense of pride" (Office of the National Cultural Commission 1998). Similarly, countless national and local exhibitions, demonstrations, seminars, festivals, and development projects focus on preserving and promoting the distinctive qualities of Thai boxing, cuisine, music, dance, art, folktales, handicrafts, medical science, silk production, weaving, and herbal medicine. The need to revitalize Thai wisdom and local wisdom was formally ensconced in the National Education Act of 1999.

Attention to Thai wisdom escalated during the period when Western influences and global economics were intensifying. As a social and educational agenda, the focus on Thai wisdom turns the gaze inward, as if there were a necessary process in developing the country outward. I refer to Thai wisdom as both a process and a product of illuminating, revitalizing, and transforming cultural knowledge and indigenous practices in ways that enable people to be more strategic in negotiating the homogenizing currents of globalization. Thai wisdom, as incorporated in the education reforms, translates into a rallying call for affirming diversity, strengthening communities, enhancing local voice, and developing more locally relevant curricula in schools.

As I have observed the phenomenon of Thai wisdom escalate in recent years, I resonate with the observations of Mebrahtu, Crossley and Johnson (2000) that "amidst this intensification of globalization . . . many communities are ever more forcefully acknowledging their distinctive characteristics and celebrating their cultural differences." I agree with them that "'cultural forces' deserve greater attention if we are to understand and deal creatively with the impact of globalization." My interest is in understanding how reformers are appropriating local wisdom in the interest of informing local curricula, and conversely, in understanding how local curriculum development processes are transforming local wisdom. For, this dynamic interaction is intended to strengthen the local voice in the interest of *balancing* official school knowledge within the context of globalization.

THE GLOBAL AND THE LOCAL

Commenting on the interaction of global and local influences, Thai philosopher and researcher Soraj Hongladarom claims that "neither absolute homogeneity nor absolute diversity is likely when the global and the local are being dynamically negotiated. Instead, the local is permeated by the global to the extent that the local finds from the global what is useful, and employs various strategies to retain its identity" (Hongladarom 2000:9). Based on his case study of how Thai youth are using an Internet café in Bangkok, Hongladarom concludes that they are far from being passively acculturated by the Western-dominated Internet. His study describes how "Thai culture co-opts the Internet" and how "identity is being constantly negotiated" (Hongladarom 2000:1). Furthermore:

The global . . . finds itself tempered by various locals to such an extent that eventually the global can no longer claim identity with any local. It is not that one local can claim globality at the expense of the other locals; on the contrary, the global that is emerging from the (ideological, political, economic, culture, social) interaction among the world's societies and cultures is such that it eventually contains elements from everywhere, but belongs to nowhere. The global, thus, becomes what I previously called "cosmopolitan." . . . It represents a makeshift mode of interacting among social entities compelled by the logic of advanced capitalism. (2000:9)

I submit that recourse to Thai wisdom and local wisdom is a strategic process in the construction of the "cosmopolitan."

Glocalization, a term coined by Roland Robertson (Featherstone, Lash, and Robertson 1997), is an apt concept here because it references the interplay between globalizing and localizing forces. This is an interplay between "both local idiosyncratic histories [and] broad global trends" (LeTendre et al. 2001), or as Robertson describes it, "the simultaneity, the co-presence of both universalizing and particularizing tendencies" (Robertson 1997). I suggest that reform projects engaged in revitalizing Thai wisdom and increasing local participation and local curricula in schools are strategic sites because they deliberately occasion glocalization, where indigenization and homogenization are both at play.

THE RESEARCH

I have studied both public and private schooling in Thailand, with Thai colleagues, since 1985 (Potisook et al. 1993; Jungck et al. 1995; Leesatayakun 1998). I was in Thailand for the year (1999–2000) that these educational reform issues were a national priority and the National Education Act of 1999 was passed. Prior to the passage of the Act I observed several innovative research and development projects that I now realize were foreshadowing these reforms, some begun as early as 1997. Five of these early reform projects have been described elsewhere (Jungck 2001). I have continued to follow several of these early initiatives as they live on through continued funding and revised versions, at new sites.

When I returned for two months in 2001 to learn more about reform initiatives, I was struck by the heightened attention to Thai wisdom and local wisdom. In the following, I draw examples from three recent specific reform projects, two led by Kajornsin and one by Thongthew, summarized in Table 1.1. The projects illustrate what is involved in honoring local wisdom and developing a more locally relevant and empowering curriculum.

Kajorsin and Thongthew are university-based educational reformers and researchers who have led multiyear collaborative reform projects in many different local communities. Collectively, all the research and reform projects summarized focus on coming to understand local communities, systems, processes, and capacities relevant to developing (versus implementing) participatory and student-centered reform goals in a variety of contexts.

LOCAL WISDOM AND
LOCAL CURRICULUM DEVELOPMENT

It is the interplay between these two that brings both into clearer focus: Local curriculum development is informed by understanding and appropriating local wisdom, while local wisdom is understood and revitalized through local efforts to appropriate and transform it through a new curriculum. Both local wisdom and curriculum are transformed in the process—or at least that is the intent. One way to conceptualize this is as Lukens-Bull (2001) does in speaking of modernity and tradition in an education context. He claims that "imagining and (re)inventing modernity is necessarily linked to imagining and (re)inventing tradition. Imagining the two is the first step in the (re)invention of both" (2001:368). For example, through developing a more locally relevant curriculum, one that introduces the concepts of sustainability and integrated farming into a traditional rice-growing community based on mono-agriculture, local wisdom is at once affirmed and re-invented or *transformed*. A new local curriculum is "imagined." However, developing local curriculum through honoring and revitalizing local wisdom and garnering community participation represents an enormous shift in Thai education policy, a shift that makes the processes of glocalization both explicit and imperative. How is local wisdom more locally defined in these projects?

Local Wisdom as Knowledge and Understanding

Reformers have defined local wisdom in ways specific to their field sites and experiences. For example, one university educator and reform project leader describes what local wisdom means to her in reference to the first project in Table 1.1:

> Three years ago, when I worked in Ban Ta Wai, they produced this kind of artistic woodcarving as their local product and they are very proud of it, very much. I stayed in that village, rented a house so that I could learn from them. And there are several families that make woodcarvings, not

just one family, not just one factory. One family produces the shape, another family polishes, another family paints, another family sells it. So, a lot of families work on this product and they hand down to each other. . . . In the village, it is like a big family and you cannot miss one family: if you miss one family, it means the product will not come out like this. So, I think that this is the local wisdom. The way they cling together, you cannot find that in Bangkok—they have to work together. It's the old tradition: Because we are an agricultural community, we have to help each other, we need manpower. (S. Thongthew, interview, August 7, 2001)

Another educational leader, Kajornsin, who is on her second multiyear reform project aimed at developing the capacities and processes through which local participants can produce local curriculum (Project 3 in Table 1.1), describes it this way:

I think that local wisdom is a person, somebody, a group of people who know something *exactly.* As they know, you know, in their daily life. Like at Ratchaburi, they are a dairy-farming community. So, we have many, many farmers who [have] dairy-farmed since a long, long time ago. So, they are experts in terms of dairy farming. Those people who know how to do, how to take care of their dairy farms, and those who are successful and got national awards, so those people who know *exactly* about the things they do in their occupations—in my opinion, those people have local wisdom. (B. Kajornsin, interview, August 15, 2001)

At one level, these reformers conceptualize local wisdom as expert knowledge, skills, and practical behaviors. For them, local wisdom is a system of meanings and understandings:

I look at this basket. . . . [it] has no meaning in itself; we look at the people who make the basket, it has some purpose. . . . Why [did] they make this shape? Why not another? This is difficult [to learn], we have to be in that place for some time before you realize why they have to make it the way it is. (S. Thongthew, interview, August 7, 2001)

Local Wisdom as Reflection and Change

At another level, however, local wisdom is seen as the capacity to reflect and understand change, as Thongthew explains:

[A] lot of people believe that culture is something that is handed down to us and . . . we must keep it . . . protect it, and hand it down to the people

Table 1.1 Reform Projects Cited

Projects and Leaders/Informants	Reform Goals Addressed	Project Focus
Project 1 S. Thongthew, 1999	Decentralization Local curriculum development Incorporating local wisdom in school curriculum Collaboration between teachers and local community Local participation	Sites: 2 villages in northern and northeastern Thailand (a woodcarving community, a basket-weaving community) 20-month study, 1999–2001, of rural philosophy (local wisdom) and how it can be integrated into school curriculum Collaboration among teachers, community members, and university educators/researchers to study the local communities and develop a curriculum based on community expertise, values, and future goals. Developed community-based curricular units, curriculum for a new local economy
Project 2 B. Kajornsin, J. Chuaratanaphong, T. Siengluecha, 1999	Decentralization Cooperation of schools and community Local curriculum development Student-centered learning Incorporating local wisdom into school curriculum	Site: 2 dairy farms in central Thailand, 2 schools, 5th grades 3½-year study, 1998–2002, to understand and develop local curriculum development processes Collaboration among teachers and their students, local dairy farmers, provincial/district/school administrators, veterinarians and dairy extension agencies, community leaders, parents, and university educators/researchers Developed, taught, evaluated, revised, and retaught new curricula using an action research model

(continues)

Table 1.1 *(continued)*

Projects and Leaders/Informants	Reform Goals Addressed	Project Focus
Project 3 B. Kajornsin, J. Chuaratanaphong, P. Potisook (unpublished)	Decentralization Local curriculum development Local participation Collaboration processes Incorporating local wisdom into school curriculum	Sites: Two villages in traditional rice- and vegetable-growing communities in central Thailand. One elementary school and one combined elementary/lower secondary 2½ year study, 1999–2002, to understand local curriculum development processes through exploration in differing contexts Collaboration among teachers, provincial/district/school administrators and personnel, and university educators/researchers Developed, taught, evaluated, revised, and retaught new semester-long curricula using an action research model

in that area. [From] another side, we believe that culture is something that is going on—okay, handed down to us from ancestors, whatever, but it has been adjusted and moves on, changes all the time, it is not the culture like in the past. So, for educators, who believe in this stance, culture is what is going on. . . . When we look at the curriculum, or local culture, we look at it the same way, we try to see the culture. . . . We do not know what had been before, but we try to collect the artifacts, talk with people in the area—especially the older people—look at ancient books. . . . We compare that culture and see the way the people in the area are now behaving. Wow, how do they change in that way? It's difficult to find out about culture in this second sense. We try to see wisdom, that is, the way they change: That's the wisdom, not just the material. We call it expressive culture. (S. Thongthew, interview, August 7, 2001)

The range of conceptions of local wisdom as I have heard and inferred them from reformers is broad and encompasses two major ways that anthropologists typically view culture: as a system of symbols and meanings and as practical activity having expressive or "performative" aspects (Sewell 1999). The latter conception is similar to Ann Swidler's notion that culture is like a "tool kit" containing "strategies of action" (Swidler 1984 in Sewell 1999).

Just as reform leaders in these projects had various conceptions of local wisdom, they also had a variety of strategies for incorporating it into reforms and, in the process, transforming it. Space does not allow me to describe these multiyear projects in detail; however, I will summarize some relevant parameters they all had in common.

Incorporating and Transforming Local Wisdom

In the first stages of the projects I observed, university-based team members spent anywhere from 6 to 18 months living in a community and interviewing community members, local school administrators, and teachers in order to study the local culture, resources, and ambitions before they and the community agreed to commence a curriculum development project together. In the next stage, teams of teachers, community members, and district curriculum administrators were assembled to work with the university-based educators. During this stage, experts (local and nonlocal) with experience pertinent to what was emerging as a desirable curricular focus were drawn into the project as well. For several months, team members would then educate themselves in specific areas (university reformers studied local wisdom, teachers studied the processes of curriculum development, community members learned how to participate, and all studied what was necessary for the new curricular content). Next, teams spent several months (often weekends at a time) jointly developing curriculum plans. After a new curriculum was developed it was taught for a semester, evaluated by all participants, revised, and taught again for another semester with a new class—such was the cycle.

The National Education Act of 1999 stipulates that up to 20 percent of a school's curriculum can be locally determined, so these projects represented only a portion of a school's total program. Though voluntary, these projects posed heavy demands on local school communities and teachers, who by tradition have not been expected, taught, resourced, or rewarded for developing and assessing their own curriculum. Similarly, neither Ministry officials nor project leaders have experience in studying or utilizing local wisdom as a basis for developing more relevant curricula. These projects are small in terms of national scope: Each involved one or two schools, one or two grade levels, and only particular curricular areas. They are huge projects, however, in terms of labor intensity, involvement of local participants, multiyear timelines, and the demands on all team members to learn new roles and embrace professional development. Most importantly, these projects foreshadow what is involved when there is a huge philosophical and organizational shift in terms of who develops and what counts as legitimate knowledge and curriculum. If revitalizing local wisdom and occasioning community participation are the means through which decentralization and glocalization proceed, then these projects are extremely important in terms of creating and understanding these new processes.

WISDOM, WISDOM EVERYWHERE, BUT "WHERE'S THE CURRICULUM?"

After studying a local community, project participants then had to figure out how to relate and transform what they learned in the communities to a specifiable curricular plan of action. I observed a variety of strategies for translating local wisdom into new and locally valued curricula. For example, Sumlee Thongthew viewed local wisdom as ensconced in the community and thus came to view the "community as the curriculum." She got local teachers (many of whom were not from the local area) to study the community much like ethnographers. They in turn developed a curriculum in which their students studied the *community's* curriculum by becoming apprentices to local artisans—in this case, woodcrafters (Thongthew 1999).

In Kajornsin's first project (number 2 in Table 1.1), located in a dairy-farming community (Kajornsin, Chuaratanaphong, and Siengluecha 1999), the schools' regular curriculum structure remained in tact, but the content was revised to be more thematically integrated around various aspects of dairy farming. Students were learning math, science, and modern technologies through visiting local dairy cooperatives, model farms, cattle breeding centers, and veterinary clinics where they were taught principles of genetics, cattle breeding, milking techniques, and animal care. While the traditional school structures and subjects were retained, curricular content was revised to infuse local wisdom, community experts and modern technologies.

Therefore, while reformers were learning ways of interacting with local communities and constructing relevant curricula within them, they were doing so in very different ways. All the projects I observed, however, were consciously tapping into local traditions, values, and practices while simultaneously introducing new currents and possibilities for both curricular and community development.

"Why Do We Have to Learn What We Already Know?"

Interestingly, a common curricular tension and philosophical question would surface at some point in all these projects. Local community members, after having been voluntarily engaged for some time in developing a local curriculum, would inevitably begin to wonder, "Why are we doing this?" Though not in these words, they began to ask one of the perennial questions in curriculum studies: What knowledge is of most worth? For example, Kajornsin explains that in her dairy-farming project, participants began to question, "Why do we have to teach dairy farming? Students, they know since they are born, they just know how to milk the cows. . . ." (B. Kajornsin, interview, August 15, 2001). Thongthew describes how this same tension and question emerged in her project:

> First of all, they go along with our local curriculum, they are enthusiastic, they work and just want children to learn, this is something new, and then, after a semester went by some of them decided they didn't want it, they said, "Why should we want our children to become like us? We want them to become doctors, someone else." (S. Thongthew, interview, August 7, 2001)

Inevitably, community members would reach a point in these projects when they questioned the value and purpose of a new curriculum based on local wisdom. Implicitly, they were wondering if this was a curriculum of social reproduction when what they wanted was a curriculum of social mobility, something more empowering. While the agriculturalists and artisans in these communities understood that their traditional work was socially valuable, they also understood that they earned little for their labor, worked in difficult conditions, had low social status. Moreover, their children wanted something different. In contrast, the government curriculum, although reputed to be irrelevant and alienating, was nonetheless seen as high-status knowledge, as it was aligned with national tests, university admissions, desired credentials, and social status. Hence the question, "Why teach local wisdom?"

If revitalizing local wisdom and developing local curricula are intended to empower people to become more "strategic" globally and nationally, then these acts have to do more than honor and reproduce local wisdom. Leaders in all these projects believed that locally relevant curricula *could* be sociologically, psychologically, philosophically, and pedagogically more empowering than the state curriculum. In fact, Sumlee Thongthew, who recently completed a 20-month project in a northern village specializing in woodcarving, saw the traditional curriculum as the cause of, not the solution to, the socioeconomic problems in rural areas. The development of a more relevant local curriculum would, she claimed, "let [the villagers] use their own wisdom and become strong in their own way" (S. Thongthew, interview, August 7, 2001):

> The equilibrium with nature and local wisdom in rural schools in Thailand has been weakened by the imposition of the National Curriculum for more than a century. The National Curriculum, much influenced by urban industrialized philosophy . . . has less regard for the long-term relationships of people with the natural environment and local culture. . . .
>
> It damages rural socioeconomic background and neglects the worthy purpose of community sustainable development. Growing migration from the rural areas to big cities emerges drastically as most graduates from rural schools find they do not have enough skills and lack attitudes necessary to work and live with nature in the rural areas.
>
> A local curriculum emphasizing the study of rural philosophy to cultivate practices and knowledge will lead to . . . sustainable development in a village. . . . The constructed curriculum highlights relevant and meaningful subject matters to the learners, the provision of learners' control over their own learning process, and community feedback of the curriculum outcomes. (Thongthew 1999:1,4)

Speaking of her experiences in her first multiyear project in the dairy-farming community, Kajornsin saw the cognitive and pedagogical values of a locally relevant curriculum that engaged students in authentic ways of "learning how to learn." For her, involving local dairy farmers and scientists from agricultural agencies in the curriculum served as a way to both utilize and enhance local wisdom while engaging teachers and students in a more relevant and constructivist, versus a memorization-based, pedagogy:

> The students were taken out of schools to study. . . . the demonstration dairy farms and local organizations became "laboratories" for information gathering. Therefore, students experienced "authentic learning."
>
> By the end of [the] second semester, we found that most students had developed more responsibility, better work habits, more expressive ability and more creative thinking. . . . The students' ability to work as a team had improved, they could present what they learned . . . they could construct their own knowledge. (Kajornsin et al. 1999:8)

There are many ways in which participants in these projects believed that what they were doing was psychologically, sociologically, and pedagogically empowering, enabling local communities to better understand and assert themselves. However, implicit in all these projects was the belief that local economic revitalization was necessary as well. In both global and local terms, these projects were expected to "add value" to communities and contribute to their economic viability and independence as well.

Transforming Local Wisdom

For Thongthew, one "value added" revitalizing economic component in the woodcarving village was the development of a local curriculum to prepare students to be local tourist guides in a new "junior hospitality business." The plan was that students would apprentice themselves to local woodcrafters in order to develop a deeper understanding and appreciation for the local wisdom, and gain "work-study" experience as well. Ultimately, they would become knowledgeable, local tour guides, employable in tourism, one of Thailand's largest growing industries (Thongthew 1999). By recasting their old and respected indigenous woodcarving village as a tourist site, the community was at once reinventing their tradition and inventing a form of modernity as well.

Kajornsin has recently completed the community study and curriculum-development stages for a new project located in a traditional rice-growing community. Community members, local agricultural experts, teachers, and district academic supervisors worked for several months to develop a new fifth-grade curriculum unit focused on "integrated and sustainable" agriculture. Students and teachers will have three rai (about 1.2 acres) of land in which to learn about integrating ecologically compatible fish farms and vegetable gardens with their traditional rice fields. The new curricular project reflects a revitalization of the traditional Thai/Buddhist values of sustainability and environmentalism, values that had been marginalized with the introduction of modern monoculture. The economic value added to the community will be derived through increased productivity enabled by a more ecologically sustainable agriculture, and new natural food niche markets aimed at health-conscious consumers. While integrated and sustainable agricultural values are consistent with the Western science of ecology and are one form of modernity, a globally distributed, capital-intensive, and mechanized mono-agriculture has characterized the modernity of many poor countries. By revitalizing its local wisdom and tradition, this community is inventing a sustainable modernity upon which it can thrive.

DISCUSSION

Thailand has introduced a comprehensive and ambitious national education reform. The plan to decentralize academic and administrative powers to local levels represents a major shift in the philosophy, politics, and practices of education. Some recent projects have focused on revitalizing local wisdom and helping local communities create processes for developing locally relevant curricula. Schools and teachers seem fully capable of the kind of reflection and action that characterizes revitalization when given the purpose and resources to do so. Moreover, glocalization occurs, for example, when communities like these revitalize their craft knowledge and re-create it in a world-class tourism center, or when poor rice farmers shift to sustainable, integrated agriculture and market their healthy "brown rice" in global networks.

At present, the links among revitalizing local wisdom, reforming local curricula, recovering the economy, and negotiating various global forces are weak in Thailand. As Thongthew notes about the prospects for educational change, "the new Bill does not provide financial resources that enable rural school teachers to accomplish the mission" (Thongthew 1999:11). Furthermore, schools are limited institutions and in order to revitalize rural philosophies and local wisdoms, authority, power, and economic resources must be decentralized as well. Some recent initiatives introduced by the new Thai Rak Thai government of Thaksin Shinawatra, aimed at rural development, move in this direction. However, the momentum of the Ministry of Education's leadership in educational reform has slowed down because of political disagreements over the nature, processes, and speed of decentralizing power. This gives more credence to political economist and social analyst Pasuk Phongpaichit's assertion that the greatest potential for economic recovery and educational reform rests not with the government, but within civil society and the local-level momentum of people's movements and NGOs (Phongpaichit and Baker 2001). Thus, these local education projects are critical sites, with high stakes, that occasion and reflect the interplay of various local wisdoms and various global forces. These projects reveal not only how long, labor intensive, and fragile these reforms are, but that the intent is for varied and nonstandardized outcomes. These Thai reformers are being more strategic and more "glocal" about globalization by supporting the local communities to revitalize their own cultural and economic resources in a significant way.

In terms of the larger questions raised in this volume, the Thai example is an interesting case. As a condition of emerging from the economic crash of 1997, Thailand adopted the ideologies and policies of neo-liberal economics and applied much of this to its education reform. Is Thailand moving toward a globalized form of schooling, validating the hypothesis of educational isomorphism? In some ways "yes," and in other important ways "no." Certainly, its new education policies reflect an appropriation of the trends toward more decentralization, the use of national standards, and managerial rationality. Like most countries, Thailand is responding to similar global pressures and education reform trends. But, these trends are interpreted and accommodated into local histories, contexts, and contemporary priorities that change over time and vary tremendously within Thailand.

As Thailand moves slowly toward a more decentralized education system, and as it increasingly and strategically strives to honor notions of local wisdom, I have seen a deliberate movement away from the kinds of conformity and isomorphism suggested by world culture theory. Variation and not conformity has been the goal and, to date, the outcome of some early reform projects. Furthermore, internal and external isomorphism is seen as one of the causes of both economic and educational problems in Thailand, something these projects are strategically poised to resist and avoid.

ACKNOWLEDGMENT

I deeply appreciate the generous support of the Fulbright Scholar Program whose 1992 research award enabled me to develop a relationship with researchers at Kasetsart University (Faculty of Education, Center for Research on Teaching and Teacher Education) in Bangkok, Thailand. Since then, Kasetsart and National-Louis Universities have consistently provided the institutional support needed for us to sustain ongoing collaborative projects focused on schools, culture and change. The views expressed in this article, however, are those of the author.

REFERENCES

Bowornwathana, B. 2000. Governance reform in Thailand: Questionable assumptions, uncertain outcomes. *Governance: An International Journal of Policy and Administration* 13 (3): 393–408.

Cheng, Y. C. 2001. Education reforms in Hong Kong: Challenges, strategies and international implications. Paper presented at the First International Forum on Education Reform: Experiences of Selected Countries, August 2, Bangkok, Thailand.

The Economist, May 24, 1997, 15.

Featherstone, M., S. Lash, and R. Robertson, eds. 1997. *Global modernities.* London: Sage.

Gosum, S. 1999. Community and school participation for basic organization. Paper presented at the National Conference on Learning Reform: A New Dimension for Human Potential Development, September 16, Kasetsart University, Bangkok, Thailand.

Hallak, J. 2000. Globalisation and its impact on education. In *Globalisation, educational transformation and societies in transition,* edited by T. Mebrahtu, C. M. and D. Johnson. Oxford, UK: Symposium Books.

Hongladarom, S. 2000. Negotiating the global and the local: How Thai culture coopts the Internet. *First Monday* 5:8 [Accessed July 25, 2001]. Available from http://www.firstmonday.dk/issues/issue5_8/hongladarom/.

IMF. International Monetary Fund's Statement of August 22, 1997. Structural reforms essential to strategy. *Bangkok Post* [Accessed 2001]. Available from http://www.bangkokpost.com.

Jungck, S. 2001. Thailand's National Education Act of 1999: Politics and contradictions of participatory reform. Paper presented at the annual meeting of the American Educational Research Association, April 14, Seattle, WA.

Jungck, S., P. Potisook, M. Leesatayakun, and S. Srisukprasert. 1995. *In the shadow of the state: A private school survives.* Bangkok: The Center for Research on Teaching and Teacher Education, Kasetsart University.

Kaewdang, R. 2001a. *Indigenous knowledge for a learning society.* Office of the National Education Commission 2000. [Accessed July 23, 2001]. Available from http://www.drrung.com.

———. 2001b. *Learning for the new century.* Office of the National Education Commission 1999. [Accessed 2000]. Available from http://www.onec.go.th/move/news/dec_16.htm.

Kajornsin, B., J. Chuaratanaphong, and T. Siengluecha. 1999. Developing a process for local curriculum development through school-community collaboration. Paper presented at Reforming Learning, Curriculum and Pedagogy: Innovative Visions for the New Century, September 15, Bangkok, Thailand.

Kwak, Byong-Sun. 2001. Education reform: How to make it a national agenda— with focus on Korean experiences on responses to the challenges of globalization and neo-liberalism. Paper presented at the First International Forum on Education Reform: Experiences of Selected Countries, August 1, Bangkok, Thailand.

Lee, J. 2001. School reform initiatives as balancing acts: Policy variation and educational convergence among Japan, Korea, England and the United States. *Education Policy Analysis Archives* 9 (13):1–14.

Leesatayakun, M., P. Potisook, P., and S. Jungck. 1998. The international program: A focus on teachers' and students' cultural perceptions and instructional strategies. Bangkok: Kasetsart University.

LeTendre, G., D. Baker, M. Akiba, B. Goesling, and A. Wiseman. 2001. Teachers' work: Institutional isomorphism and cultural variation in U.S., Germany, and Japan. *Educational Researcher* 30 (6): 3–15.

Lukens-Bull, R. 2001. Two sides of the same coin: Modernity and tradition in Islamic education in Indonesia. *Anthropology and Education Quarterly* 32 (3): 350–372.

Madrick, J. 2001. There's no one plan that fits all. *Bangkok Post,* August 6, 8.

Mebrahtu, T., D. Crossley, and D. Johnson, eds. 2000. *Globalisation, educational transformation and societies in transition.* Oxford, UK: Symposium Books.

Ma-oon, R., P. Srisukvatananan. 1999. Research and development of a cooperative learning model of whole school reform in Chiang Mai. Paper presented at the National Conference on Learning Reform: A New Dimension for Human Potential Development, September 15, Kasetsart University, Bangkok, Thailand.

The Nation (Bangkok). 2001a. Kingdom: lesson in development. July 31, 5A.

———. 2001b. World Bank urges reform. July 23, 10B.

Neher, C. D. n.d. *Contemporary Thailand.* Northern Illinois University [Accessed 2001]. Available from http://www.seasite.niu.edu/crossroads/cneher/cn.thailand.thm.

Office of the National Cultural Commission. 1998. *Culture campaign against economic crisis.* Office of the National Cultural Commission. [Accessed 2001]. Available from http://www.culture.go.th/english/CultureCampaignAgri.htm.

ONEC. Office of the National Education Commission. 1999. National Education Act of B.E. 2542 (1999). Bangkok, Thailand: Office of the Prime Minister.

———. 2001. The first international forum on education reform: Experiences in selected countries: Conference proceedings brochure. Bangkok, Thailand: Office of National Education Commission.

Pascoe, S. 2001. Education reform at school level: From policy to practice. Paper presented at the First International Forum on Education Reform: Experiences in Selected Countries, July 30, Bangkok, Thailand.

Phongpaichit, P., and C. Baker. 1998. *Thailand's boom and bust.* Bangkok, Thailand: Silkworm Books.

———. 2001. Modern society, community culture, or theme park: Four debates on the future. Paper presented at the Council on Thai Studies, November 2, Northern Illinois University, DeKalb, IL.

Potisook, P., M. Leesatayakun, S. Srisukprasert, and S. Jungck. 1993. *In the shadow of the mosque: Secular schooling in context.* Bangkok, Thailand: Center for Research on Teaching and Teacher Education, Kasetsart University.

Prawase, W. 1998. *Reform of Thai society: A national agenda. Bangkok Post.* [Accessed 1998]. Available from http://www.bangkokpost.com/.

Redmond, M. 1999. *Wondering into Thai culture.* Bangkok: Redmondian Insight Enterprises Co., Ltd.

Robertson, R. 1997. Comments on the "global triad" and "glocalization." Paper presented at the Globalization and Indigenous Culture Conference, May 15, Kokugakuin University, Japan.

Sewell Jr., W. H. 1999. The concept(s) of culture. In *Beyond the cultural turn: New directions in the study of society and culture,* edited by V. E. Bonnell and L. Hunt. Berkeley and Los Angeles: University of California Press.

Srisalab, N., and W. Sritongsuk. 1999. *Thai wisdom and education.* Radio Thailand FM 95.5, WWW. Available from http://www.onec.go.th/move.news/aug99/aug14.htm

Srisukvatananan, P., and R. Ma-oon. 1999. Research and development of a cooperative learning model of whole school reform in Sakhon Nakhon. Paper presented at the National Conference on Learning Reform: A New Dimension for Human Potential Development, September 15, Kasetsart University, Bangkok, Thailand.

Supradith Na Ayudhya, C. 2000. From crisis to adaptation: A model for educational service area offices. Melbourne, Australia: University of Melbourne.

Swidler, A. 1984. Culture in action: Symbols and strategies. *American Sociological Review* 51: 273–286.

Teachers of Baan Khnong Tao and Baan Tung Laung School. 1999. Research and development of [a] student-centered learning model: A case of Baan Khnong Tao and Baan Tung Laung School. Paper presented at the National Conference on Learning Reform: A New Dimension for Potential Development, September 16, Kasetsart University, Bangkok, Thailand.

Thongthew, S. 1999. The return of rural philosophy in Thailand. Paper presented at the 16th annual Conference Exploring New Frontiers in Education, November 20, Hong Kong.

Watson, K. 2000. Globalisation, educational reform and language policy in transitional societies. In *Globalisation, educational transformation and societies in transition,* edited by T. Mebrahtu, M. Crossley, and D. Johnson. Oxford, UK: Symposium Books.

Zack, M. 1997. A question of class. *Far Eastern Economic Review 160* (49): 48.

CHAPTER TWO

TRANSFORMATIONS IN SOUTH AFRICA
Policies and Practices from Ministry to Classroom

Diane Brook Napier

South Africa is one of many nations embarking on democratization and modernization of its education system. A sweeping national reform program reflects South Africa's adoption of global ideas including decentralization, democratization, and "outcomes-based education" (OBE). However, educational reform is creolized and re-creolized (Hannerz 1987) as it moves through South Africa's system from national policy to provincial mandate and from provincial mandate to local and classroom practice, as redirection or reinterpretations occur. The difference between reform and reality, the policy practice dichotomy, is a phenomenon examined in research on reform and implementation issues in an array of developing countries that adopted ideas from outside (for example, Arnove 1999; Brook Napier in press; Paulston 1976; Reimers and McGinn 1997; Riddell 1998; Rust and Yogev 1994; Samoff 1999a and 1999b). For real understanding of educational reform, we need to consider the variety of lived cultures of schooling in South Africa and other countries, as well as the viability of world culture theory regarding educational reform. What factors shaped the creolized and re-creolized versions of reform at intermediate and local levels in the South African system, and what influences were at the heart of the imported reforms to begin with? Considering the large picture, is South Africa joining a world system of education? And in the narrower view, looking inside the country, to what degree is South Africa experiencing the emergence of a new homegrown hybridized, variously creolized, system of education?

THE GLOBAL-LOCAL MODEL
APPLIED TO SOUTH AFRICA

In a highly centralized system like South Africa's, reforms usually originate at the national level, where decision makers have often been influenced by global trends in educational reform. Reform ideas then "cascade" down to the next level in the hierarchy, and the next, and the next, until they finally reach teachers and administrators in schools. However, instead of simply trickling down, at every level ideas can be transformed and sometimes blocked. Further, international consultants and trainers as well as foreign-trained South Africans can introduce ideas from outside at any level in the system, particularly given the new principles of devolution of authority and decentralization. Figure 2.1 represents the South African reform process and serves as the framework for the discussion in this chapter, showing how the cascade model works in South Africa. At the top is the global to national transition, where features of educational reform in many different countries—the ingredients in the world culture theory of reform—appear in the South African version of plans for educational transformation. Here are the broad brushstrokes of reform policy: democratization, deracialization, decentralization. The reforms also embody a "paradigm shift" away from the traditional to progressive, learner-focused forms of curriculum, teaching, and evaluation, which South Africans often refer to collectively as OBE, outcomes-based education.

Level 1 in Figure 2.1 represents the national or macro level of reform, South Africa's own version of the global ingredients that have been packaged in a series of plans, initiatives, and programs. The national reforms are implemented at the provincial level (Level 2), where some degree of modification or creolization is bound to occur because of "devolution of authority" to the provinces and because of the differences among the provinces. Another intermediate level exists in the form of specific activities of training, support, and oversight at the subprovincial level (Level 3), where further re-creolization is prevalent. Finally, at the micro level (Level 4) of schools and communities, further re-creolization can occur, as teachers, administrators, and other local actors sometimes resist, mediate, and transform the substance of the reforms into forms shaped by internal realities and contextual factors.

In this chapter, after an overview of some global influences evident in South Africa's reforms, I consider the national reforms that are then translated

Figure 2.1 Global-Local Model

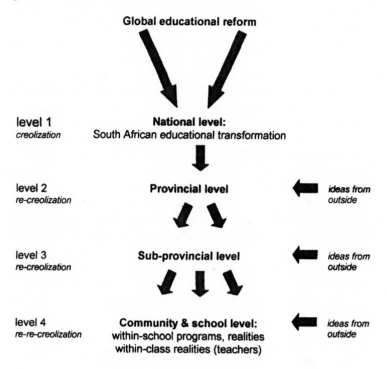

into provincial mandates (Levels 1 and 2). Thereafter, I focus more specifically on what happens at Levels 3 and 4, the subprovincial level and the school/community level, both of which represent critical interfaces where re-creolization occurs. To illustrate how these processes can happen, I offer examples from the past decade of change and reform. The examples shed light on the implementation issues that have become a growing focus of the reform debate, as well as on the manner in which reform ideas can be creolized as a result of selective adoption, reinterpretation, resistance, or other processes.

The examples of Level 3, training and support activities, come from published reports of training programs and my own recollections from training programs in which I was the presenter alone or with South African colleagues in the period 1995–2001. To illustrate some of the realities in schools and communities at Level 4, and how these can influence the manner in which reforms are entertained, I offer examples from micro-level research in schools across

South Africa in the last decade, including my own field research in schools and archival research conducted in South Africa in the period 1991–2002 in consort with a variety of training and staff development activities. The examples are descriptive studies: surveys of teachers' perceptions of their needs and of the effectiveness of training and support programs; and some ethnographic studies in schools with data obtained from participant-heavy observation, interviewing, and examination of curriculum materials. These studies expose a range of issues and dilemmas inherent in implementing the various programs that constitute South Africa's educational transformation plan. They shed light on consistent cries from teachers about their anxieties and frustrations, and on the need for coordinated training and support.

Following these accounts of South African educators grappling with the implementation of reforms, I present a series of lessons that might be learned, as an elaboration of the kinds of transformations that have occurred, and of the factors determining the nature of impact of the reforms as they are translated from policy into practice. Finally, I conclude with some broader implications based on the illustrations and on the possible lessons learned. What can one surmise from the past decade of experience of reform as policy (in various forms and phases) and of reform as practice? What might one expect to see happen as South Africa proceeds into the next phases of its sweeping reform agenda for education?

GLOBAL INFLUENCES

Some ideas from abroad were adopted intact or nearly so by the architects of the plan for transforming South Africa's highly centralized, traditional, externally evaluated complex of racially segregated school system—with its dominant-culture, Eurocentric curriculum—into a single, nonracial, democratic, relatively decentralized system. In the broadest terms, South Africa joined a worldwide trend of reform to democratize its education system, to make it less discriminatory. The blueprint for South Africa's reform was developed with significant infusions of ideas from the United States: for instance, on competency-based teacher education, outcomes-based education, performance-based evaluation and assessment systems, and learner-centered pedagogy. The U.S. practice of pegging teacher training, performance, and outcomes to a series of "standards," measured by systems of "indicators" (as in the NCATE system for accrediting programs in colleges of teacher education), was adopted wholesale into the reform programs. Similarly, South Africa's new national program of teacher education emphasized "competence-based" teacher education (e.g., SCE 1996) as opposed to the traditional knowledge-based approach, borrowing schemes of "competences" from programs in England (themselves influenced by American ideas) and intermingling them with the ideas borrowed from the United States. (See the discussion of training under Level 3 for an illustration.) The structure of South Africa's new National Qualifications Framework, an integrated framework for learning achievements at all levels, was based on the frameworks introduced in Britain, Canada, and New Zealand in the 1990s (ANC 1994; South Africa 1997). However, South Africa paid little attention to the reforms and experiences in neighboring Botswana and Zimbabwe, particularly Zimbabwe, where a similarly entrenched apartheid type of education system had been dismantled after independence (Lemon 1995).

NATIONAL REFORM POLICIES

The overall transformation agenda in education was part of the multisector transformation agenda (often referred to as the "new dispensation") under the new government, wherein education was seen as one of the crucial human resource development areas along with health, housing, jobs, and water supply. The national educational reforms came in a series of waves and stages that I review in brief here.

The apartheid era of racially segregated education officially ended in 1994 with the installation of a multiracial democratic government, and the adoption of a new Constitution in 1996. However, educational transformation had been initiated earlier, in a 1987 white paper and in the 1991 Educational Renewal Strategy (South Africa 1991).

The national reforms contained several major thrusts. Quantitative or structural reforms aimed to dismantle the racially segregated system, as one of the most significant thrusts was for democratization and deracialization. Racially segregated education had been one of the so-called "pillars of apartheid." Four separate education systems (for whites, Coloureds, Indians/Asians, and Africans) had had vastly different levels of funding and resources and separate schools and teacher training colleges. There had been different education departments for each of the four systems in each province as well as separate departments for the "homelands" (South Africa 1992). Education for Africans had been by far the most disadvantaged, administered by the Department of Education and Training (DET). In the New South Africa, a deracialized system was envisioned, one that was still centralized at the national and provincial levels, but one that gave communities and schools new autonomy to implement the reforms locally and that gave teachers new responsibilities and roles. The DET was abolished; the four systems were collapsed into one system; and all public schools became "multiracial." The latter changes alone amounted to a significant step in transformation, a break with the racist past. For an overview of the policy context and the magnitude of the challenges facing South Africa in education, see Brook (1996 1997) and Jansen and Christie (1999).

The process of "rationalisation" or downsizing of the system was necessary to eliminate duplication of institutions and facilities that had existed under apartheid. This feature of South African reform was somewhat unique, required to dismantle the apartheid-era structures, but it resembled the process that had occurred in neighboring Zimbabwe, formerly Southern Rhodesia, which had had a similarly complex form of apartheid education. The rigid, top-down system was to be decentralized by devolution of authority to the provincial and local levels. This feature resembled reforms for decentralization in many countries with colonial-era centralized systems of education. Another thrust was to modernize the system and make it more globally competitive, for instance in emphasizing mathematics, science, and technology as priority subjects, addressing labor force needs for skilled workers, and elevating levels of literacy and numeracy among South African schoolchildren and adults.

Qualitative reforms of curriculum, teaching, and learning were also initiated, in the creation of the National Qualifications Framework and adoption of the principles of OBE. Under these rubrics, South Africa established programs to reform pre-service teacher training and in-service training and support; curricular content and textbooks; and systems of quality assurance evaluation and certification, teacher professionalism, and administration. Special initiatives addressed key areas of need, including mathematics, science, literacy, and special education. Underlying all of the reforms was the notion of a paradigm shift that demanded a new way of thinking about and operating in education. Rather than examine any one reform in detail, I provide a general overview. In illustrations later in the chapter, I focus on a selection of these OBE reforms that impacted teacher trainers, teachers, and administrators as a total package, a series of mandates and initiatives that were intertwined. Detailed accounts of the importation of OBE from abroad and the implications for South African reform can be found in Jansen and Christie (1999).

The National Educational Policy Act of 1996 articulated the major provisions for democratized education based on constitutional principles. The South African Schools Act of 1996 articulated transitional arrangements including devolution of (some) authority to schools, governing bodies, and communities; racial integration of all schools; new roles and responsibilities for teachers; and abolition of corporal punishment of pupils. The controversial Curriculum 2005 outcomes-based reform program introduced in 1998 produced a torrent of criticism (for example, Chisholm 2000; Cross et al. 1998; Cross and Rouhani 2001; Jansen and Christie 1999; Potenza 2000; Review of Curriculum 2005 2000; and South Africa 2002). In 1999, the Ministry of Education responded to the criticisms of its policies by declaring that there was a crisis in all subsectors of education and announcing an ambitious five-year plan to redirect the reforms under the slogan of "Tirisano" [working together] to address the implementation problems and to improve teacher training, support, and oversight as well as coordination across levels of administration (South Africa 2002). In 2000, Minister Asmal announced an external evaluation plan for all schools, the National Framework for Whole School Evaluation, under which every school in the country would be evaluated and goals for improvement in dysfunctional schools would be set (SAIRR 2002b). In 2001, a National Curriculum Statement was released, announcing a refined version of Curriculum 2005 for compulsory education through grade nine to be implemented between 2004 and 2008. Minister Asmal declared that under this new curriculum "no pupil would leave school at the end of grade nine unable to read, write, count, and think to a high level" (SAIRR 2002b:278).

Alongside criticisms of reforms, there were calls for investigation of implementation realities (for instance Brook 1996, 1997; Cross and Rouhani 2001; Cross et al. 1998; Herbert 2001; Howard and Herman 1998; Jansen 1997; Jansen and Christie 1999; and Motala 2001). National reforms and initiatives were one story; how they played out within the provinces and in local communities was a different story, and one demanding scrutiny.

PROVINCIAL LEVEL MANDATES, DIFFERENCES IN CONTEXT

At Level 2 in Figure 2.1, authority is devolved to the provinces to implement the reforms and the provinces can develop their own mandates and initiatives as regional policies. Since each province has its own administrative structure, mission, and way of operating, one might expect differences from province to province in the implementation of national reforms. In other words, some re-creolization is inevitable, particularly since it falls to the provinces to enact and oversee the key processes of implementation, including training, support, and oversight.

A variety of factors contributes to the possibilities for some re-creolization of reform at this level. South Africa has nine provinces (Figure 2.2), significantly different in degree of development and prosperity. For example, Gauteng (containing the cities of Pretoria and Johannesburg) and Western Cape (containing Cape Town) are the wealthiest provinces, boasting the nation's highest adult literacy rates (98.1 percent and 95.8 percent, respectively, in 1996), highest matriculation rates (74 percent and 83 percent in 2001), lowest proportions of underqualified teachers (11 percent in each in 2000), and the lowest poverty rates (28 percent and 20 percent in 2001) (SAIRR 2002a). Because they have the country's leading universities, the highest density of other colleges and technikons (technical colleges), and NGOs, they offer the richest variety of options for training and support experiences. However, they have large numbers of disadvantaged schools and large impoverished populations in informal settlements and townships. In contrast, Limpopo and Eastern Cape have higher poverty rates (exceeding 60 percent in 2001), and fewer facilities for educational training and support. Provinces also differ in population size, language use, rural/urban development, available resources, legacies of inequities in education ("backlogs" in disadvantaged former nonwhite schools), in-school realities, provincial politics and personalities, and local community attitudes. Hence, within the provinces, at the subprovincial Level 3 in Figure 2.1, there is no level playing field for implementing any of the reforms, either across provinces or within provinces.

Case 1: A Province Redefines Modernization

As an example of a national-level initiative transformed by a province, consider Gauteng Province's response to the nation's Educational Technology Forum. When the national ministry proposed that there be at least one computer in every school, Gauteng Province in turn announced an ambitious and expensive plan to link the province's schools to the Internet and equip each school with computers, although in practice the policy ran into trouble right away (SAIRR 2002b:279; J. van der Vyver, personal communication, June 2001).

In the next section, I discuss the variety of training experiences within the provinces during the last decade to illustrate how complex the process of implementing reforms becomes in practice and how this leads to a variety of outcomes.

FROM PROVINCIAL MANDATES TO SUBPROVINCIAL TRAINING, SUPPORT, AND OVERSIGHT

Within each province, one sees the first critical interface between reform as policy and reform as practice. Teachers, teacher trainers, and other actors

Figure 2.2 Provincial Map of South Africa

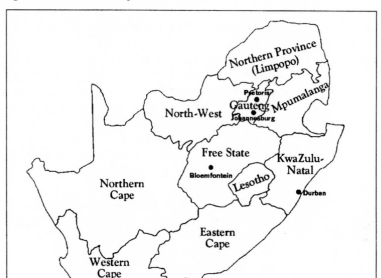

are often first confronted with the implementation of the reform mandates in the context of training and staff development programs. A myriad of institutions and organizations within each province participates in this process, with varied outcomes. The training and support component of the reforms is crucial since teachers and administrators at all levels have a heritage of training and experience in a system characterized by external examinations and curricula, top-down decision making, and little room for initiative.

People conducting the training can be master teachers in school, members of provincial education departments (often called learning facilitators, or LFs), or faculty in universities or teacher training colleges. Members of NGOs and foreign consultants often bring expertise on imported ideas such as standards for curriculum development and evaluation. Large budget development projects funded by donor agencies and international organizations add to the mix (cf. Samoff 1999a). Training endeavors are often partnerships between several of these groups or individuals, involving possible competing agendas. Access to training is unevenly distributed across the country and within provinces, which creates still further complication. Various training and support programs encompass the different major components of the national reforms and provincial mandates. The emphasis on training in a particular component of the reforms has also shifted periodically, and teachers have not received training in all aspects of reform they are required to implement. Programs vary in duration from hours to several weeks, and many training programs are one-shot undertakings with no follow-up. There are many corners of the country where educators in schools have little or no access to any training or support activities. Educators in the historically disadvantaged institutions are particularly impacted by this (Brook and van der Vyver 1998).

The challenges of moving from policy to practice through training and support activities are illustrated in these cases with examples from my own experiences with teachers and teacher trainers at the subprovincial level, in teacher training colleges, and in schools. How have teachers fared as they engage with the new ways of operating and thinking, and what were the outcomes of training programs? Here, we can ponder how training endeavors played out in some cases, as an illustration of how imported ideas entered South African reforms, found their way into the system as national mandates, and then penetrated teacher training colleges or schools as the focus of training programs.

One thrust of OBE reforms revolved around the need for developing systems of quality assurance (QA) in teacher training colleges and schools. QA elements were incorporated into the national plan for revamping teacher education, first in the transitional years (1994–1999) as "competences" for teacher education that were laid out in the National Qualifications Framework policy documents and others, including the *Norms and Standards for Teacher Education* (South Africa 1997). From 1998 onward, the focus shifted slightly to meeting "strategic objectives" for addressing standards of performance (competence) that were subdivided into a series of indicators. Preliminary versions of these were to be used as the basis for developing internal QA systems. Under the principle of devolution of authority, the specifics of developing internal systems were not explicated, leaving the provinces and institutions within them free to develop their own plans. The examples that follow show how the process of transformation in these cases of QA was reinterpreted, truncated, or selectively implemented, depending on internal factors at the subprovincial level.

Case 2: A Workshop Leads to No Quality Assurance Plan

In staff development workshops lasting three or four days apiece and conducted annually from 1996 to 1998, my colleague J. van der Vyver and I worked with several groups of 20 to 50 faculty from two teacher training colleges and several schools in Gauteng Province.[1] The workshops were part of a multiple-year partnership between the provincial department, the teacher training colleges, several local suburban and township schools with which the college faculty worked, and visiting scholars from the United States. They were designed to train the staff in ways to address the "professional competences for teacher education" (SCE 1996:3). An internal document introduced the idea of a competence-based approach, saying, "the national programme does not indicate the exact skills or range of skills that a teacher should acquire while at college, but does suggest a few guidelines" (SCE 1996:3). For structuring the training experiences in the two colleges, a set of dimensions of competence ("understanding children's learning, management of learning, planning for learning, teaching skills, and personal and professional development") was adopted intact from a program in a college of education in Lancashire, England. Here is an example of ideas from abroad entering the system at the subprovincial level in response to national mandates and skipping over the provincial level.

The workshop participants told us that they liked the consistency of having the same workshop presenters every year, people that they knew, and they appreciated our use of English and Afrikaans interchangeably. Positive personal relationships facilitated our training efforts. Only one faculty member refused to participate; the other participants (roughly 150 in total) tackled the workshops with seriousness.

Some concerns and resistance emerged, too, in the workshops. The participants were suspicious of the imported "competences," seeing them as alien and unnecessary. Several asked why we could not "just give [them] a syllabus or instructions" so that they could operate in the style to which they were accustomed. Schoolteacher participants commented that a workshop on tackling the realities in their schools (such as absenteeism and violence) would be more valuable than these competences "from the Department" (they perceived the competences as coming from the national level); but if they were to live up to these competences, they said they wanted supervision to be sure they were "doing it properly."

Participants from the two teachers colleges admitted they were more concerned about rumors that the college would be closed the next year than about the substance of the workshops. Why did they need the workshops if the college would be closed and they lost their jobs? Their concerns were justified. Both colleges were closed in a subsequent year, when the number of teacher training colleges in Gauteng was reduced by two thirds under mandates for downsizing. When we suggested that having experience in professional competencies and performance standards would benefit the participants even if they moved to another job, our argument sounded hollow.

What was the actual outcome of these training activities? No formal plan or document was developed for either college. Although the workshop participants engaged with the new ideas quite intensely, and the competence-based approach became part of their discussions of work, the impact was truncated by closure of both colleges. Some staff were reassigned to posts in other colleges; perhaps there they continued their learning and adoption of the new ideas but we are speculating on this. Many staff at these colleges left the education sector when the colleges were closed.

Case 3: A Quality Assurance Plan Aborted

In 1998, 1999, and 2000, I was one of five presenters in another series of three-day workshops focused on helping participants develop their own plans for QA, now guided by a set of strategic objectives developed at the provincial level. The participants were approximately 200 faculty and administrators at a large distance-education college in Gauteng. The ultimate goal was to develop a "Culture of Quality Assurance" at the college and an internal plan for QA. The workshop team consisted of presenters from an NGO and the college itself and consultants from the provincial department and abroad. Each year, the participants bravely tackled this task of developing a quality assurance system for their college, agreeing with the administrators that documentation of the college's efforts to address mandates was for their own protection. Each year, however, we, the presenters, were asked if we could not first comment on whether or not the college would be closed the following year. By the third year of workshops, anxieties over the threat of closure made the presentations increasingly difficult, with participants worried about the likelihood of being redeployed to other posts or being offered retrenchment packages. "Will we be here next year?" became a standard question. Welch (1999) described the complex task involved in this kind of training, when participants' enthusiasm for receiving the new ideas was in competition with concerns over job security and the future.

What was the outcome at this college? The QA workshops did produce an internal document, an explicated QA plan for individual self-assessment of heads of departments, senior heads of departments, and department programs. We held a formal ceremony marking the release of this document and the culmination of a productive cycle of training experiences. This was supposed to be the first step, followed by subsequent cycles in which the QA plan would be developed further. However, here too, the process was truncated by closure of the college when its programs were absorbed into the University of South Africa (UNISA), the country's leading distance-education institution, under the national mandates for restructuring of higher and further education.

Case 4: Workshops That Bring the First News of Quality Assurance

In a similar series of quality assurance workshops in the southeastern Free State in 1998–99, I again heard concerns about job security from participants, who were 40 to 60 faculty from the teachers college and the local university, a historically disadvantaged institution (HDI) formerly for Africans. These participants were outspoken that the grim realities of rural schools in their area (isolation, no books, classes of 60 to 90 pupils, low teacher pay) were more important to address than the new ideas from the province or the national government. Many participants said they had not seen the policy mandates before, nor the performance standards and indicators in the materials we were using in the workshop. They identified ambiguous wording in some performance indicators and were skeptical of using these preliminary versions of the QA guidelines from the provincial department, wondering if there would be a different final version. Details did change in the QA guidelines after the first year and this created resistance to the workshops in the second and third years. The participants also complained about inadequately qualified learning facilitators from the provincial department who had done other workshops here. They expressed frustration with contradictions in what was emphasized from one workshop or presenter to another, and with the endless round of new initiatives. "What next?" was a frequent question posed to us. Some commented negatively on "outside experts" from universities in South Africa or abroad, who knew little of the local realities (cf. Brook and van der Vyver 1997; Brook Napier, Lebeta, and Zungu 2000).

In this third QA example, the outcome of the training experience was similar to that in the first example I described, where no formal plan of QA was developed for the college. The workshop participants did become minimally acquainted with features of the new system; but here too the process was truncated when the teachers college and the HDI were absorbed into the University of the Free State, the province's leading (formerly white) university.

Transformations of the Message

The cases of QA workshops reveal that training experiences designed for implementation of reforms had varied outcomes and impacts. While it was evident that some acceptance of the new ideas did occur, the original message of the training programs was somewhat creolized by various factors. These included participants' skepticism and begrudging acceptance of the ideas and loyalty to traditional ways; frustrations of working with preliminary versions of the QA guidelines, which were rushed into use; and wariness of inadequately qualified learning facilitators and "outside experts." Closure of the colleges certainly interrupted the original intent of the training, which was to be implemented through the level of the teacher training colleges to schools.

These examples are a small slice of the totality of experiences that exist at the subprovincial level in South Africa. However, they contain insights into the precarious process of translating reform mandates into local practice. They also show how difficult it is to implement multiple sets of reforms at the same time, when these processes can conflict with one another and when reform is fast-paced. At the micro level of communities and schools (Level 4 in Figure 2.1), one can contemplate the possibility of an even greater variety of realities and experiences. Examples of research on realities in schools and communities are presented below.

REALITIES IN SCHOOLS AND COMMUNITIES

At the local level, teachers and other actors in schools adopt, mediate, resist, or reject reforms in keeping with various influences and contextual factors. Here is another critical interface where reform as practice occurs. As teachers and administrators in schools resist change or enact change to one degree or another, they influence the degree of penetration of reforms to the school level. The need for more research on local realities has become increasingly evident as implementation issues provoke debate over the transformation process. The cases that follow offer us insights into school realities as South Africa's education system has undergone the early years of transformation. These examples come from my own research as well as that of other scholars in South Africa. For additional examples, not described here, see Howard and Herman (1998) and Jansen and Christie (1999).

Case 5: Developing Local Forms of Deracialized Schooling

Some research showcased innovative schools that changed ahead of official reform. Beginning in the mid-1970s and continuing into the early 1990s, "open" (progressive, private, nonracial) schools became models of how all schools might become multiracial and democratic. These pioneer schools represented grassroots involvement in transforming South African education. They were exceptional cases of bottom-up reform rather than the usual top-down cascading reform process. The challenges in these schools later became more widespread as the entire system underwent transformation after 1994. In open schools, teachers needed training or retraining to use multiple forms of materials in teaching, to deal with multiple languages in the classroom, to take the initiative in developing curriculum with a multicultural focus, and to deal with students from vastly different racial, cultural, and socioeconomic backgrounds. The innovative programs in these schools were shining examples of the "right kind of transformation," as it were. They developed indigenized, contextualized programs that incorporated the principles of democratized, multiracial schooling. They developed the first indigenized curricula in the country—in local versions of British Integrated Studies, imported from England and including English, history, geography, and science, as well as in art and music. Teachers confronted the challenges of teaching content alongside teaching English to non–English speakers, and they participated in governance and program development. The crucial role of leadership was evident, in that many of these schools had an ethos and a supportive administration that were conducive to adopting new ways (Brook 1991, 1996, 1997; Brook Napier in press; Christie 1990; Freer 1991; and McGurk 1990). What kind of transformation was evident in these schools? They epitomized educational transformation and creolization of reform ideas into a locally contextualized and indigenized new form of South African schooling true to the spirit of the overall reform plan.

Case 6: Transformations in Other Schools

A variety of other transformations can be shaped by in-school realities and local contextual factors, including attitudes and perceptions of teachers. Surveys and qualitative research provide illustrations here. In a National Schools Register of Needs Survey, fewer schools were declared "unfit for education" in 2000 than in 1996, but significant numbers were still woefully deficient in facilities. For example, 17 percent lacked toilets, 34 percent lacked running water, and over 70 percent had no computers, testifying to the persistent backlogs that have plagued African schools in particular (SAIRR 2002b:274).

In qualitative research in a variety of school settings, I described how "blocking factors" impeded reform, perpetuated disadvantage, or significantly shaped in-school realities (Brook 1996). These factors included geographic location, money, language, resistance to change within institutions, inadequate teacher in-service programs, the cycle of school dropouts and unemployment, and persistent apartheid-era backlogs particularly in African schools. Well-resourced suburban schools were vastly different from impoverished township schools a few miles away. Rural

African schools were the most disadvantaged, having few options for improving their situation regardless of reforms. In my study, five such schools had experienced virtually no change. Even well-resourced mission schools with strong programs were disadvantaged by isolation. The principal of a mission school in far northeastern KwaZulu Natal lamented, "I never have a chance to hear new ideas from other principals. . . . we are so far away" (Manzini 1994 quoted in Brook 1996: 215).

Several qualitative research studies investigated teachers' perceptions of their training experiences with OBE, with implications for transformation. Fleisch and Potenza (1998) reported that although teachers in a Gauteng study felt pressured to change their practices in the context of Curriculum 2005 directives, training largely in short courses with little practical value or feedback resulted in little change in teachers' practices. They reported on a subsequent more teacher-friendly project piloted in Gauteng. In a study of OBE training programs for grade one teachers in Northern Province (Limpopo) schools, Khumalo et al. (1999) also reported mixed effectiveness of the workshops and teachers' expressed need for better training and support. Luneta (2001) reported similar findings in surveying attitudes of teachers, teacher trainers, and student teachers in Mpumalanga Province. In a study of 158 primary school teachers in North West Province, Pitsoe and Niewenhuis (2001) found that many teachers were skeptical of OBE; and they identified underresourced schools and insufficient in-service training (particularly in follow-up, practical implementation of OBE principles) as key problems. Smit (2001) also pointed to the importance of considering teachers' opinions of OBE training effectiveness, their confidence, and their attitudes about professionalism.

School-level practices vis-à-vis reforms were noted by van Vollenhoven (2001), whose research illustrated how the mismatch between educational practice and legislative provision—in this case the Code of Conduct for Learners in the Schools Act of 1996—creates difficulties in transforming particular aspects of schooling. Finally, in a case study of a school in the Western Cape, Jansen (1998) argued that the way schools' governing bodies use the powers accorded them under the Schools Act can also influence the implementation of transformative reforms.

Case 7: Thwarting Deracialization

Implementing legislation for racially integrating all schools was not always smooth sailing. Most integration of government schools has been one-way, of nonwhite pupils into former white schools. In some former white government schools, the apartheid-era school ethos and attitudes remained unchanged, particularly in the transitional years immediately after 1994. Brook (1996) noted examples of suburban Johannesburg schools in which, at that time, the only real change in response to reforms was in the admission of African pupils; they were segregated from white pupils within classes. Similarly, Odav and Ndandani (1998) reported their study in schools in two Afrikaans farm communities, one in Northern Province (Limpopo) and one in Northwest Province, where there was violent resistance against racially integrating the former white schools as mandated by the South African Schools Act of 1996. The measures to racially integrate these schools produced protracted in-school segregation. In one of these schools, black pupils were placed in a practically separate subschool, taught in English, and on a schedule separate from the Afrikaans-medium main school, which remained overwhelmingly white. In both cases, protests and violent resistance erupted among whites and blacks, as both groups were antagonized over the manner of implementing the mandate to integrate. Odav and Ndandani suggested several lessons from the two cases: that historical factors (such as conservative rural community values) need specific attention when reforms are implemented; that transitional arrangements for integrating and reforming schools (as per the Schools Act) need to be "tightly arranged around the issue of mechanisms to resolve the problems of integration, all white governing bodies, dominantly white schools, (and) Afrikaner language schools"; that students need to learn discipline in resistance; that the media can distort the picture by failing to report key aspects; and that there had not been any "leveling of the playing field" at the civic and school levels (1988:8–9).

Transformations

As illustrated in these accounts of research in South African schools, what kinds of transformations are evident? I have already suggested that the open schools represent the desirable kind of transformation, in which creolization of reforms results in locally contextualized, indigenized programs that essentially run true to the spirit of the overall plans for transforming education. However, in most of the other examples given here, focusing on in-school realities and responses to reforms, a variety of different transformations emerges. In remote-area and disadvantaged schools (Brook 1996), exclusion from change persists because blocking factors persist. In the cases reported by Fleisch and Potenza (1998), Jansen (1998), Luneta (2001), Khumalo et al. (1999), Pitsoe and Niewenhuis (2001), Smit (2001), and van Vollenhoven (2001), one sees a variety of examples of selective enactment of change: some resistance and teacher concerns mixed with acceptance of reforms and participation, and issues arising from relationships between in-school practice and legislative provision. In these illustrations, it is hard to ascertain how much the original spirit of the reforms is retained.

The issue of racially integrating schools, in the examples of research in former white schools reported by Brook (1996) and by Odav and Ndandani (1998), focuses on one of the most fundamental tenets of the whole reform plan, to deracialize the nation's schools. In the research cited, nonwhite students were indeed admitted to former white government schools, but how this occurred essentially subverted the original intent. In-school or in-class segregation amounted to getting around the rules, mangling the intent of the mandate to integrate in the wrong kind of transformation. Here, re-creolization can be interpreted as token integration only. Loyalty to traditional ways, selective adoption of reforms mixed with resistance, and the power of local contextual factors combine here to suggest that transformation and creolization processes are complex indeed.

This sampling of research on realities in schools suggests that there is fairly widespread awareness of OBE and other reforms, but implementation issues include high levels of need, anxiety, and frustration among the teachers as they confront the reform mandates. Moreover, the precise nature of change at the school and community level is widely variable and unpredictable. What is predictable and consistent are calls for better trickle down of reforms through more effective training and support programs and through better understanding of how official reforms run up against local realities in practice. There is ample evidence of change, but whether or not it is change true to the spirit and conception of South Africa's version of democratized education and OBE remains open for further debate.

LESSONS LEARNED

What lessons can we learn from these insights and experiences? What kinds of transformations have been occurring in South Africa and with what impact? Here, we suggest five lessons.

Penetration varying with context.   A rich mosaic of different manifestations of transformation is developing in different regions and places, with different versions colored by local contextual factors including apartheid-era legacies related to race, language, ideology, resources, school type (and setting and location), the nature of the training and support experiences and follow-up (if any), local and provincial politics, and leadership.

Cynicism but not (usually) outright rejection.   Skepticism and anxiety among educators, but also willing participation in reform, are entwined threads in the cases presented here. Educator uncertainty was focused on job security concerns with good reason, as I have demonstrated. When official guidelines and training materials are rushed into use via poorly coordinated training, resistance is bound to follow. It is understandable if teachers lack confidence to operate under the new paradigm. Partial resistance, selective enactment of change, and loyalty to traditional ways emerge in many cases. Outright rejection of reforms can point to important contextual factors and the need for better transitional arrangements, as argued by Odav and Ndandani (1998).

Mixed impact in subsequent years. Implicit in the workshop experiences and the research was the notion of consistency, or lack of it. In the first year of a multiple-year workshop cycle, the novelty was evident; in subsequent years contradictions and inconsistency combined with growing job security concerns to dilute participants' enthusiasm and the impact of many workshops. Changing priorities handed down as ever more initiatives and mandates can wear teachers down. Success in one year does not guarantee sustained success.

Timing and pacing of reform. It was widely considered that the government was trying to rush OBE. The pace of implementing reforms has been bruising for actors at lower levels in the system, yet foot-dragging would perhaps be worse. We witnessed inertia in many teacher training colleges where we presented programs: The staff were so focused on anxiety about their future that proactive behaviors regarding reform ideas were rare. Even workshop duration and timing created a conundrum for us: Participants burned out after three days, yet there was so much new to present and learn. Reforms are being implemented in a short span of years, on top of a decades-long legacy that resides in people's minds. What timing and pacing are ideal?

Changing focus, priorities. The number of changes and the official claims made about each new initiative have prompted skepticism about the overall process. Perhaps the lesson here is that radical and sweeping reforms are bound to provoke opposition, no matter how well conceived they are. If policy can learn, as Jansen (1997) asked, it should learn from experiences at lower levels in the system, where it is creolized and re-creolized in response to provincial and local realities and contextual factors.

CONCLUSION

Micro-level research in schools across South Africa and experiences in training programs suggest that as nationwide transformation continues, a mosaic of different manifestations of transformation is developing in different regions and places. There is a patchwork of intended outcomes (such as emerging operationalization of some OBE practice), but also many unintended outcomes within and across schools and in communities. Vertical and horizontal discordance in implementation, the pacing of reform initiatives, and frequent changes in priorities have generated anxiety and frustration among educators.

What are the actual outcomes to date? Nowhere in South Africa is change completely absent. Even for the most disadvantaged and remote schools, the law of the land has changed. Officially, there is a single education system with new forms of funding, legislative recourse, and opportunity for equitable participation. At the other extreme, has there been wholesale transformation, to the letter of the law and reforms, or as an exact duplicate of imported ideas? Nowhere have these occurred either, since under the principles of devolution of authority and decentralization the process is open-ended and the outcomes are divergent. The most valid assessment of outcomes to date might be articulated as some degree of transformed schooling everywhere. In some cases, change might still be primarily in policy rather than in practice. Where change is seen in practice in schools, it can take any number of forms along a continuum of hybrids, from situations in which creolization equates to "desirably contextualized schooling" in the spirit of the national plan (as in the open schools), to situations in which creolization is a mangled version of the original intent (as in examples of token integration of former white schools), and with a variety of other permutations in between these extremes.

What does this mean for South African transformation in general? In the broadest sense, one might argue that South Africa is indeed joining a world system of education in developing a democratized, somewhat decentralized, nonracial or multiracial system of more progressive education, with emphasis on priorities such as mathematics, science, technology, literacy, and multiple languages. Within this, to what extent is South Africa experiencing the emergence of a homegrown, creolized, or hybridized form of education? The Department of National Education and the Provincial Education Departments have creolized imported ideas for democratized schooling and OBE into a sophisticated system of reforms and mandates. At the subprovincial level, the complex variety of training experiences adds a layer of complication and re-creolization. Locally, teachers and administrators in schools are engaged in either compliance, mediation, or resistance, or some combination thereof. As they, too, re-creolize the mandates based on their own needs and perspectives, they generate varieties of forms within and across schools even in the same community. A basic contradiction emerges, that transformative democratizing reforms encounter centralized mindsets shaped under apartheid both in the mechanisms for implementation and in their reception by teachers and administrators, many of whom still want oversight and structure as they are accustomed to. There is also the dilemma that new problems are being added on top of persisting old problems because there are so many unpredictable, varied, and inadequately documented manifestations of OBE.

On the positive side, the examples offered in this chapter point to widespread participation in reform as practice, and there is hope for the continuing emergence of a transformed, better system of education. In a survey of South Africans' views of the most serious problems not resolved since 1994, education ranked fifth (17 percent of respondents) after unemployment, crime and violence, housing and shelter, water, and sanitation. Educational differences were identified as the dominant cause of inequality, not race or poverty (SAIRR 2001). Public recognition of the value of education might well provide continued support for the reforms.

Considerations of the viability of a global template for educational reform clearly need to include the creolization and re-creolization processes within countries. Based on the experiences in South Africa, the global-local model oversimplifies the picture. Other layers are needed, reflecting the provincial level and the subprovincial level. Any global template should include consideration of school-by-school and within-school variations based on local, internal, and personal legacy factors. Timing and pacing are other important variables demanding long-term studies to understand what is stable versus what is fleeting. We can contemplate the educational transformation process in South Africa in three dimensions: vertically, as change trickles down from top to bottom and somewhat from bottom up; horizontally, as change unfolds across the country in different provinces, regions, and places; and longitudinally over time, to ascertain if early progress was sustained in progressive schools, if and as new change emerges in other schools, and as change in many forms continues across all schools. South African transformation is likely to attract international attention for some time to come. Within the country, we need to learn more about the complex landscape of transformed communities and schools, including the success stories, the disasters, and the rich variety of cases in between these extremes in an entire system that is in flux.

NOTES

1. John van der Vyver was Senior Lecturer at Soweto College of Education, then served as a senior administrator in the Gauteng Department of Education until his death in 2002. He developed and oversaw many programs to train

teachers and college of education faculty in competence-based approaches and in QA.

REFERENCES

ANC. African National Congress. 1994. *A policy framework for education and training*. Johannesburg: ANC Education Department.

Arnove, R. F. 1999. Reframing comparative education: The dialectic of the global and the local. In *Comparative education: The dialectic of the global and the local*, edited by R. F. Arnove and C. A. Torres. Oxford: Rowman and Littlefield.

Brook, D. L. 1991. Social studies for multicultural education: A case study of a racially integrated school in South Africa. *Georgia Social Science Journal* 22 (1): 1–10.

———. 1996. From exclusion to inclusion: Racial politics and educational reform in South Africa. *Anthropology and Education Quarterly* 27 (2): 204–231.

———. 1997. South Africa after Apartheid: Recent events and future prospects. *Social Education,* Special Issue on Sub-Saharan Africa, 61 (7): 395–403.

Brook, D. L., and van der Vyver, J. 1997. *Partnerships: Highways and byways.* South African Association for Research and Development in Higher Education, Johannesburg, March.

———. 1998. Policy implementation issues in Historically Disadvantaged Institutions. Paper presented at Tenth World Congress of Comparative Education Societies, Cape Town, July.

Brook Napier, D. (In press). Language issues in South Africa: Education, identity, and democratization. In *Language and Inequality,* edited by R. Terborg and P. Ryan. Mexico City: UNAM Press.

Brook Napier, D., T. V. Lebeta, and B. Zungu. 2000. Race, history, and education: South African perspectives on the struggle for democracy. *Theory and Research in Social Education* 28 (3): 445–451.

Chisholm, L. 2000. *Report on Curriculum 2005.* Pretoria: Department of National Education.

Christie, P. 1990. *Open schools: Racially mixed Catholic schools in South Africa 1976–1986.* Johannesburg: Ravan Press.

Cross, M., and S. Rouhani. 2001. Curriculum reform in South African Basic Education: A paradigm shift? Paper presented at the 45th Annual Meeting of the Comparative and International Education Society, Washington D.C., March.

Cross, M., Z. Mkwanazi-Twala, and G. Klein, eds. 1998. *Dealing with diversity in South African education: A debate on the politics of a national curriculum.* Johannesburg: Juta, Kenwyn.

Fleisch, B., and E. Potenza, 1998. School based teacher development: Preliminary Findings from a South African Pilot. Paper presented at the Tenth World Congress of Comparative Education Societies, Cape Town, July.

Freer, D., ed. 1991. *Towards open schools: Possibilities and realities for non-racial education in South Africa.* Manzini: McMillan Boleswa.

Hannerz, U. 1987. The world in creolisation. *Africa* 57: 546–559.

Herbert, R. 2001. Lost in translation: Outcomes based education in South Africa. Paper presented at the annual meeting of the American Anthropological Association, Washington D.C., November 29.

Howard, S., and H. Herman, eds. 1998. Coping with rapid change: Special focus on South Africa's teachers. *Democracy and Education* 12 (2): 1–49.

Jansen, J. 1997. Can policy learn? Reflections on "Why OBE will fail." In *Outcomes education and the new curriculum.* University of the Witwatersrand Educational Policy Unit, Education Seminar Series, unpublished document.

Jansen, J. 1998. Grove Primary: Power, privilege, and the law in South African Education. Paper presented at the Tenth World Congress of Comparative Education Societies, Cape Town, July.

Jansen, J., and P. Christie, eds. 1999. *Changing curriculum: Studies on OBE in South Africa.* Johannesburg: Juta, Kenwyn.

Khumalo, L. P., W. D. Papo, A. M. Mabitla, and J. D. Jansen. 1999. *A baseline survey on OBE in Grade 1 classrooms of the Northern Province, South Africa.* Center for Education Research, Evaluation and Policy (CEREP), University of Durban-Westville, and the Faculty of Education, University of the North, South Africa.

Lemon, A. 1995. Education in post-apartheid South Africa: Some lessons from Zimbabwe. *Comparative Education* 31 (1): 101–114.

Luneta, K. 2001. Teaching practicum: A triangulation of college lecturers', co-operating teachers' ands students' perspectives: A case of Ndebele College of Education, South Africa. Paper presented at the Ninth BOLESWA International Educational Research Symposium, Gaborone, Botswana, July-August.

McGurk, N. J. 1990. *I speak as a white: Education, culture, nation.* Marshalltown: Heinemann Southern Africa.

Motala, S. 2001. School reform in South Africa: Surviving or subverting the system? Paper presented at the 45th Annual Meeting of the Comparative and International Education Society, Washington, D.C., March.

Odav, K., and M. Ndandani. 1998. Between school policy and practice: Comparing the Potgietersrus and Vryburg crises in the light of the South African Schools Act 1996. Paper presented at the Tenth World Congress of Comparative Education Societies, Cape Town, July.

Paulston, R. G. 1976. *Evaluating educational reform: An international casebook.* ERIC document ED133243.

Pitsoe, V. J., and F. J. Nieuwenhuis. 2001. The views of teachers on the impact of OBE on classroom management: A study of schools in the Lichtenburg District. Paper presented at the Ninth BOLESWA International Educational Research Symposium, Gaborone, Botswana, July-August.

Potenza, E. 2000. No name change for C2005. *The Teacher,* September 19. Johannesburg: Daily Mail and Guardian.

Reimers, F., and N. F. McGinn. 1997. *Informed dialogue: Using research to shape education policy around the world.* London/Westport: Praeger

Review of Curriculum 2005. 2000. *The Teacher,* July 24. Johannesburg: Daily Mail and Guardian, *http://teacher.co.za/200007/curriculum_resource.html*

Riddell, A. R. 1998. Reforms of educational efficiency and quality in developing countries: An overview. *Compare* 28 (3): 277–291.

Rust, V. D., and A. Yogev, eds. 1994. *International perspectives on education and society, Volume 4, Educational reform in international perspective.* Greenwich, CT: JAI Press.

SAIRR. South African Institute of Race Relations. 2001. *Race relations and racism in everyday life.* Fast Facts 9/2001, pp.2–12. Johannesburg: South African Institute of Race Relations.

———. 2002a. *Provincial profiles: How the provinces shape up against one another.* Fast Facts, 6/2002, pp. 4–7.Johannesburg: South African Institute of Race Relations

———. 2002b. *South Africa survey: Employment, education and the economy,* pp. 3–32; *Education,* pp. 239–284. Johannesburg: South African Institute of Race Relations.

Samoff, J. 1999a. Institutionalizing international influence: The context for education reform in Africa. In *Comparative education: The dialectic of the global and the local,* edited by R. F. Arnove and C. A. Torres. Oxford: Rowman and Littlefield.

Samoff, J. 1999b. No teachers guide, no textbooks, no chairs: Contending with crisis in African education. In *Comparative education: The dialectic of the global and the local,* edited by R. F. Arnove and C.A. Torres. Oxford: Rowman and Littlefield.

SCE. Soweto College of Education. 1996. *School experience, guidelines for a competence-based approach.* Internal document, Soweto College of Education, Soweto.

Smit, B. 2001. Untitled paper presented at the 45th Annual Meeting of the Comparative and International Education Society, Washington D.C., March.

South Africa. Department of National Education. 1991. *Education renewal strategy: Discussion document.* Pretoria, South Africa: Committee of Heads of Education Departments.

———. 1992. *Education realities in South Africa.* Report, Education Policy Branch. Pretoria: DNE.

———. 1997. *Norms and standards for teacher education, training, and development.* Discussion Document, Technical Committee on the Revision of Norms and Standards for Teacher Education. Pretoria: DNE.

———. 2002. Department of Education website. http://education.pwv.gov.za/

van Vollenhoven, W. 2001.Untitled paper presented at the 45th Annual Meeting of the Comparative and International Education Society, Washington D.C., March.

Welch, T. 1999. Evaluating the quality of assessment in teacher development programmes: Criteria, methods, and findings from recent research. Paper presented at the First NADEOSA (National Association of Distance Education Organizations of South Africa) National Conference, Pretoria, August.

CHAPTER THREE

TEACHING BY THE BOOK IN GUINEA

Kathryn M. Anderson-Levitt and Boubacar Bayero Diallo

The Republic of Guinea in West Africa has a centralized educational system that controls teachers right down to daily sign-off on lesson plans by school directors. Yet, ironically, during our research on reading instruction in Guinea, we noted that the Ministry of Education was promoting massive reforms that seemed to encourage teacher autonomy. One was a project to improve teacher skills and support student-centered instruction in every elementary classroom in the country, a project supported by the United States Agency for International Development (USAID).[1] Another was the Small Grants School Improvement Project, supported by a World Bank loan, which encouraged local teachers across the country to propose school-level reforms and then compete for funding to carry them out (Diallo et al. 2001). An ambitious program of teacher recruitment and training, which had strong Canadian participation, was also framed within "a strategy of professionalizing teaching" since teachers "have to continually make professional decisions" (Diané and Grandbois 2000:8). Finally and of particular interest in this chapter, the nationwide distribution of a new set of textbooks, which was supported by a loan from the African Development Bank, had inspired Ministry of Education staff to argue that teaching methods should be "in the teacher, not in the book."

John Meyer and Francisco Ramirez suggest that there is a global movement toward "the nominally professionalized and somewhat autonomous teacher" (2000:126) as part of a larger convergence toward a global model of modern schooling. Were we witnessing Guinea's participation in a world convergence toward greater teacher professional freedom?[2]

As it turned out, the Guinean situation was not that simple. Outside of Guinea, there was not a homogeneous movement toward teacher autonomy but rather a debate between forces for teacher professionalization and forces for scripted teaching. Inside Guinea, international actors filtered the debate to transmit only pro-autonomy elements, while local decision makers appropriated selected elements of the pro-autonomy argument to suit their local purposes. Meanwhile, many teachers preferred scripted teaching, even though in practice they deviated from prior scripts the Ministry had supplied. Thus, the discourse in Guinea about teacher autonomy did not point after all to diffusion of a homogeneous world model. Rather, it revealed actors arguing for locally motivated positions and, in doing so, sometimes making use of selected elements of a cultural debate that was taking place in other countries (cf. Rosen, this volume).

TEACHER AUTONOMY

When we ask whether teachers exercise autonomy, we have in mind five interrelated questions: (1) Who sets general educational goals (and evaluates their accomplishment)? (2) Who selects the specific topics to be studied, such as which sounds to study in first-grade reading? (3) Who determines timing, that is, the sequence of topics and the pace of instruction? (4) Who selects the textbooks and other materials? and finally, (5) Who shapes the structure of lessons, that is, what happens day to day in class? Our case concerns the control of reading instruction in the earliest elementary grades, and we focus on teachers and the "state"—that is, the Ministry of Education and its deputies and advisers. We do not consider control exercised by parents or the community, an issue raised by Rosen and by Stambach in this volume, nor pressure to conform that may come from colleagues (see Paine and Ma 1993; Shimahara and Sakai 1995).

Countries vary in the degree to which the state officially attempts to control what happens in classrooms. For example, in the United States, state boards or local districts determine curricular goals, may set specific topics, and often mandate textbooks (e.g., Schwille et al. 1988). Still, U.S. teachers are relatively free, within the constraints of the texts available and of statewide testing, to determine what happens inside the classroom, and when. In France, there is even less official control. The Ministry outlines objectives for the year, but teachers select their own texts and structure lessons as they choose, with relatively little interference from the Ministry's inspectors (Alexander 2000; Anderson-Levitt 2002).

However, we are interested not just in official policy but in who actually controls classroom practice. This question has drawn considerable attention in the global North. For example, in the United States, observational studies have shown that teachers do not conform slavishly to textbooks and teachers' guides (Barr and Sadow 1989; Freeman and Porter 1989; Hoffman, et al. 1998; Stodolsky 1989). Fewer scholars have studied teacher autonomy in the global South, and the reports are contradictory. For example, Flinn describes local teachers in Pulap, Micronesia, as running classes according to their own schedule and, in the absence of books, as developing materials that "reflect their own experiences and understanding" (1992:50; see also Baker 1997 on an aboriginal teacher in Australia). In fact, some observers argue that teachers in the South exercise *too much* autonomy in the sense that they receive too few classroom visits from the headmaster, as Fuller noted in Malawi (1991), or too frequently absent themselves from school, as Baker noted in Sri Lanka (1997). Yet observers often argue that teachers in the South exercise very little freedom. In Vietnam, Baker observed that "no room is left for innovative or flexible teaching, even if a teacher were so inclined" (1997:458), and in Cuba, she witnessed "a formal curriculum . . . followed down to the last detail" (1997:462).

Scholars who see little freedom for teachers in the South have offered many hypotheses about why. First, some note, the state seeks to control teachers through symbols of modern management such as timesheets and scripts (Fuller 1991), or through inspectors and examinations (e.g., Alexander on India 2000:244; Kumar 1990; Watson-Gegeo and Gegeo 1992). Second, real or perceived lack of skills, including lack of fluency in the language of instruction (Watson-Gegeo and Gegeo 1992) or inadequate knowledge of the subject matter (Iredale 1996) might make teachers dependent on a script (Meyer 1998). Third, difficult circumstances such as lack of materials or huge numbers of students in the classroom might encourage script-driven teaching (Fuller 1991).

This chapter will consider how teachers in Guinea felt about autonomy, whether they exercised it in the classroom, and why. First, however, we will ask what counted as autonomy among different Guinean reformers and where their discourse fit within international movements for (and against) teacher professionalization.

EDUCATION AND REFORM IN GUINEA

Guinea is a nation in the midst of tremendous educational reform. Although it is one of the least schooled countries in the world (World Bank

1994:216–217), by 1997 enrollment rates had gone up to 50.5 percent of an age cohort (35.5 percent of girls) and were continuing to climb (SSP 1997:2). We focused our study on language arts and reading instruction in first- and second-grade classes, which since 1984 has taken place in French, although virtually no students speak French at home. Learning to read in French is crucial to school success.

Despite some movement toward decentralization, Guinean education has remained highly centralized. The Ministry of Pre-University Education supervised schools through a structure of regional and local inspectors. In the mid-1980s, the country adopted a national curriculum, which included a 111-page booklet governing first and second grades. The curriculum laid out the goals for reading instruction, the topics (which sounds) to be covered each year, the sequence in which to cover them, and a suggested lesson plan. Guinea's National Pedagogical Institute designed official language arts textbooks for the first and second grades, *Langage-Lecture* [Language-Reading], along with teachers' guides that provided more detailed suggestions for lesson plans. The textbooks and teachers' guides, which aligned closely with the official curriculum, replaced an old colonial-era textbook that many teachers had been using.

As mentioned, the question of professional autonomy arose in the context of several ongoing reforms in Guinea. The situation that we were able to study most closely was the distribution of a brand-new textbook for reading and language arts. By 1998, copies of *Langage-Lecture* had become scarce. During our fieldwork that year, Guinea suddenly acquired a new, French-published series entitled *Le Flamboyant* [The Flame Tree], and the Ministry approved it as a second official textbook for language arts and reading. The new textbooks were not explicitly designed to encourage teacher autonomy, but their arrival forced teachers, school directors, and local supervisors to choose between the older and the new textbooks and to manage gaps between the new textbooks and the official curriculum. What's more, the Ministry had not acquired the teacher's guide for the new series and had not yet launched anticipated training seminars, so that teachers had seen no sample of scripted lessons from the new textbook. The new textbook thus provoked a great deal of talk, from the Ministry to schools, about how much teachers should improvise when teaching reading.

RESEARCH METHODS

During the spring and fall of 1998, we conducted participant observation in 11 different schools in the capital, Conakry, in three provincial towns, and in two rural villages, visiting nine first-grade classes and nine second-grade classes.[3] Each classroom visit lasted from half a day to five full days. We interviewed the classroom teachers and many school directors and supervisors in those localities, and we collected a written questionnaire from 100 teachers and teacher trainers. We also interviewed teacher trainers and policymakers in the Ministry and the Pedagogical Institute, as well as international consultants and the educational officers of international donor agencies.[4] This paper also draws from observations in 11 elementary schools across the country conducted in 1994 for other purposes (Anderson-Levitt, Bloch, and Maiga Soumaré 1998).

COMPETING MODELS OUTSIDE GUINEA

Seen from inside Guinea, reform efforts might have looked like part of a juggernaut moving toward greater professional autonomy for teachers around the world. The reforms we mentioned at the beginning were supported by advisers from other African nations, France, England, Canada (that is, Quebec), and the United States. In fact, however, within some of the countries from which the movement for autonomy sprang there were also countermovements that advocated more government control of curriculum, planning, and the structure of lessons. When international reformers encouraged greater teacher autonomy in Guinea, they were conveying only one side of a two-way debate that was raging in countries like England and the United States.

In England—where elementary teachers had exercised tremendous freedom to control topics and their timing, what happened in class, and even some of the goals of instruction in the 1960s and 1970s—the government had more recently bound teachers to a fairly detailed national curriculum and a fairly well scripted plan for a daily literacy hour (Judge 1992; United Kingdom 1997–2002, 1999). Meanwhile, in the United States the influential reports of the Carnegie Task Force (1986) and the Holmes Group (1986) encouraged greater teacher autonomy in the context of longer professional training. Yet simultaneous U.S. reform movements sought greater control over teachers, with tests of basic competency, more detailed descriptions of their jobs, and standardized testing of their students (Judge 1992). For example, at least two of the methods for initial reading instruction recommended by the American Federation of Teachers feature scripted lesson plans (Gursky 1998; Northwest Regional Educational Laboratory 2001), and some U.S. teachers apparently accept or even embrace standardized curricula (Datnow and Castellano 2001; Goodnough 2001).[5]

We do not mean to imply that a debate about autonomy versus scripted teaching rages in *every* country. Such tension seems to exist, for example, in

Namibia (Snyder, 1998; Zeichner and Dahlström 1999) and may be built into outcomes-based education in South Africa (Herbert 2001). However, no such debate goes on in France. There, people take teacher freedom in the classroom entirely for granted (Alexander 2000; Anderson-Levitt 2002). Likewise, to the best of our knowledge, teacher autonomy is not an issue in Japan, where collegial interaction seems to produce homogeneous lesson structures; or in India, where educators do not seem to question teachers' lack of autonomy. We see the debate about control of teaching as transnational but not as "global" in the sense of "universal."

REFORMER DISCOURSES ON TEACHER AUTONOMY

Inside Guinea, there were many actors involved at different levels in the shaping of educational reform (Anderson-Levitt and Alimasi 2001). In this chapter, we will focus on Guinean decision makers and experts, on their international advisers, and on classroom teachers, with some mention of regional inspectors.

Broad Autonomy

Many international consultants expressed a desire that teachers exercise what we would call autonomy broadly defined, such that teachers determine topics, timing, textbooks, and lesson plans. For example, a French adviser in Guinea argued, "You can teach reading with a newspaper, with a print ad. It's a little condescending [*un esprit un peu infantilisant*] to say that the teacher has to follow a particular book." This expert was arguing for broad autonomy, French style, in which the state sets the goals but teachers use any text, any topics, and any classroom methods to achieve the state's objectives.

Some consultants even advocated a degree of teacher control over instructional goals, arguing that teachers should be free to "seize the teachable moment" in response to the needs or interests of their students. Such a philosophy aligned with the ideal of a more student-centered curriculum (also posited as global by Ramirez and Meyer 2002), since student-centered instruction requires teachers to be free to respond to students' needs and interests. In this spirit, one North American consultant who had observed many classrooms complained about Guinean teachers,[6]

> They are not encouraged to think for themselves. It's so predictable; exactly the same thing on the same day was what we saw. They are not encouraged *to seize the moment, the teachable moment.* They are encouraged to get

through the program, to "cover" the year. That's why we hope to encourage supervisors to encourage teachers *to be free to create.* [emphasis added]

Another consultant trained in Europe and North America expressed a similar notion when we asked about an ideal classroom situation:

> I would like to see, for instance, that a teacher could take advantage of an arithmetic lesson *to digress* into French. . . . Teachers tend to tell me, "We can't do that. It's Saturday, so we do this." . . . I think they have a margin of maneuver that they don't realize. . . . As for me, I prefer to take more time *rather than to "cover" the curriculum.* . . . So I look for teachers who are interested in their students' learning, who have a *curiosity.* . . . Teachers hesitate to *play* with ideas. [emphasis added]

Both of these consultants spoke of a creative, even playful, breaking away from the official curriculum's guidance, and both described Guinean teachers as rigid and controlled.

It is not surprising to find a French adviser advocating French-style teacher autonomy, or to find other international consultants expressing an even stronger ideal of teacher freedom. As an official from USAID explained, that agency hoped to see movement "from a teacher-centered approach to one that demonstrates equity or democracy" (see also Zeichner and Dahlström 1999:xvi), and, as we saw, student-centered instruction implied teacher freedom. Meanwhile, some of the international advisers we quoted above had worked or been trained at institutions instrumental in shaping the Holmes Group and Carnegie Task Force reports advocating professional autonomy in the United States.[7]

On the other hand, very few Guinean experts expressed support for broad professional autonomy. One Guinean did comment about teachers in a private conversation, "They know their students' levels, they know the students. They know what is appropriate for their class. Any proposition that comes from [the Ministry in] Conakry is illusory." However, this expert added, "This is just my idea, eh? Not the Ministry's."

Limited Autonomy

Indeed, the Ministry of Education did not promote teacher autonomy in a systematic way, nor had it adopted the principle of professional autonomy as part of its official policy. Only when copies of the Ministry-designed textbook became scarce did some Ministry experts begin to refer, informally, to the value of professional autonomy. When they did so, they argued for a limited autonomy in which teachers might decide which materials to use and perhaps how to present them in class, but not what topics to teach or when to teach them. As one decision maker put it,

> In the current context in Guinea, I keep saying that we must not train people to use textbook X or Y, but rather that we need to train our teachers so that, no matter what the textbook, in reading, science or other subjects, they are capable of adapting it to their teaching.

Similarly, at a training session for supervisors to prepare them for the new textbooks, we heard the slogans "It's not the book that makes the teacher" and "A new book does not mean a new teaching method." Such slogans seemed to advocate freedom from any one textbook but not necessarily from a common curriculum and teaching method.

Certain international advisers likewise preferred limited rather than broad teacher autonomy. A North American expert commented, "I don't know if I see teachers' independence from manuals [teachers' guides] as a priority." This expert reminded us that Guinean children had to take standardized exams and therefore might have less chance of succeeding if their teachers wandered from the official goals and topics.

There was a logic to Ministry experts' desire for limited autonomy. Since textbook scarcity was likely to continue, it was in the state's interest to have teachers who could operate with any materials or with none, and who did not depend on scripted guides. One Ministry official explained:

> We have received and continue to receive as gifts or as purchases . . . books that are not necessarily adapted, notably sciences books not necessarily adapted to our curriculum. But should we for that reason refuse books in a milieu where there is only one book per 10, 12, 15 students? Our teachers must be prepared to be able to adapt their books.

In this context, Ministry personnel made use of discourse that was already available from the other reform projects we have mentioned; unlike the new textbooks, talk about autonomy at training workshops did not just "fall from the sky." At the same time, the Ministry had already invested heavily in developing national goals, the sequence of topics, sample lesson plans, and national examinations. To advocate broad autonomy and hence abandon the present program would have required developing an alternative system for training teachers and assessing students—a costly affair. Moreover, national and regional authorities hesitated to cede too much of their power to local schools.[8] Thus, limited autonomy for teachers was a practical philosophy: It sanctioned the flexibility required by lack of resources without threatening the larger system.

MIXED MESSAGES FROM GUINEAN TEACHERS

According to our written questionnaire, many teachers, too, subscribed to an ideal of limited autonomy. Yet, when presented with the real dilemma of a new textbook, many insisted on the need for a scripted teacher's guide to accompany it.

Declared Preference for Limited Autonomy

One of us (Diallo) collected questionnaires from 100 educators (a 91 percent response rate) at a regional training seminar for use of the new textbook and at schools visited in the region in October 1998. Of the respondents, 20 were inspectors and teacher trainers,[9] 4 had had varied training, and the remaining 76 were school directors and teachers. The questionnaire included the following open-ended question about autonomy: "The official curriculum prescribes a certain sequence for the teaching of sounds. If a textbook does not follow that sequence, what should a teacher do?"

Only nine respondents answered this question by arguing for what we judged to be broad professional autonomy (for example, "A teacher must manage every pedagogic situation in class. He must adjust his aim"). Of these nine, most were inspectors and teacher trainers (Table 3.1). A third of the teachers and directors argued instead that when official curriculum and text conflict, teachers should exercise not autonomy but conformity—conformity either to the curriculum, to the textbook in hand, or to the advice of a superior. In between these two extremes, almost half of our respondents proposed limited autonomy for the teacher. For instance, some suggested that in the case of conflict, teachers would have to take it upon themselves to adapt the textbook or to use a different book. Others suggested that the teacher could compose the text required by the program on the blackboard, or could fill the gaps by having students do role-playing to generate the needed text.[10]

Yet Demand for a Scripted Teacher's Guide

Despite the openness to limited autonomy expressed on the questionnaire, when faced with a similar real-life dilemma, many teachers expressed the desire for very explicit guidance. Lesson layout in the brand-new textbook *Le Flamboyant* did not correspond perfectly to lesson layout in the older and now scarce *Langage-Lecture*. Moreover, its sequence of sounds did not correspond

Table 3.1 Preferences in the Hypothetical Case

Respondents	n	Broad Autonomy (%)	Limited Autonomy (%)	Conformity (%)	Ambiguous or Missing Response (%)
Inspectors and teacher trainers	20	25	50	5	20
Teachers and directors	76	5	47	34	13
Others	4	0	25	25	50
Total	100	9	47	28	16

to the sequence in the official curriculum and old textbook. "Without the teacher's guide, which we had for *Langage-Lecture, Le Flamboyant* is not easy," said one second-grade teacher in a provincial town. A supervisor elsewhere commented about the new textbook, "As it's presented [without a teacher's guide], the instructor is going to have to be a real master teacher." In an urban school, a second-grade teacher brought out the older textbook's teacher's guide to show us how, as she said, "It gives a complete lesson outline [*la demarche*]." Indeed, the old teacher's guide laid out a script for a sample lesson in detail, which was what she wanted to see for the new textbook.

We can offer several hypotheses on why teachers sought explicit guidance. First, in the absence of an official Ministry policy encouraging improvisation, teachers could rightly fear the reaction of inspectors. The memory of the authoritarian regime that ended in 1984 was fresh enough to make teachers cautious. Second, lack of confidence in the French language encouraged dependence on scripted lesson plans, for, as you will see, teaching reading required a good bank of French vocabulary. A sense of inadequacy in French was especially strong among teachers who had grown up during Guinea's 17-year experiment with instruction in national languages. Third, teachers may have felt community pressure to teach by the book (cf. Ouyang 2000). Although Guineans respected teachers who mastered more than one textbook or who could teach well no matter what the text, they also respected formality and the written word. For example, meetings and workshops in Guinea tend to be carefully orchestrated and recorded, and, in Diallo's experience, traditional Quranic instruction hews closely to a familiar script. Fourth, autonomy would have required more work from already burdened teachers. Although Guinean educators have devoted incredible effort to projects they valued (Diallo et al. 2001), they labored under circumstances that made extra effort very costly. In urban areas, teachers faced classes of 60 to 110 students and often taught double sessions. Many worked a second job to make ends meet; some struggled with attacks of sickle cell anemia or malaria. Finally, greater autonomy would have posed an equity problem for rural schools, where teachers tended to be less well prepared, books and other resources were harder to come by, multigrade classes were common, and lesson preparation was more difficult due to lack of electricity. Teachers had good reasons, then, for seeking scripted instruction (cf. Snyder 1999).

LIMITED AUTONOMY IN
ACTUAL CLASSROOM PRACTICE

Ironically, however, even as teachers asked for scripted lesson plans for the new textbook, they did not adhere religiously to the script already provided for the older textbook. Despite the impression of international observers that teachers taught "exactly the same thing on the same day," the teachers we observed exercised a little bit of leeway in their choice of textbook, moved through the curriculum at only roughly the same pace, and improvised on the proposed lesson plan.

Choice of Textbook

Although almost all the teachers we visited taught from one of the two officially approved textbooks, the exception was interesting. Among teachers observed in 1994, 14 out of 15 first-grade teachers were using the approved text and only one used the recently supplanted colonial textbook *Mamadou et Bineta*. Among teachers observed in 1998, 17 out of 18 used either the older approved textbook or the brand-new book or both, but one still used the colonial textbook. The latter case of nonconformity merits notice, for by 1998 the colonial textbook had been "forbidden" for many years. Moreover, this was a young teacher working in a large urban school where her colleagues used approved textbooks. We wondered whether more teachers used the disapproved text than our observations had revealed, particularly since a teacher who used an approved textbook herself said of the colonial textbook, "It's a good book. A lot of teachers use it, if they want to reinforce."

Sequence and Pace of Lessons

Among the teachers using the approved texts, we witnessed conformity to the officially prescribed sequence of topics, but some variation in pace of the lessons. The official curriculum provided for one lesson per week in the first grade and two lessons per week in the second grade. True to this recipe, teachers we observed almost always began a new lesson on Monday no matter what had happened in class the previous week. Even so, we found some variation in the timing of lessons in 1994. For example, during mid-March observations, the 13 teachers using the first-grade *Langage-Lecture* textbook were teaching any lesson from Lesson 16 through Lesson 24.[11] With 25 lessons in the entire book, this range of variation represented 32 percent of the year's curriculum.

In 1998, in schools that adopted the new textbook, teachers had to choose between the official sequence of sounds and the sequence presented in the new text. For instance, when teacher Madame D was pressed by her director to begin using the new textbook, she began with its lesson on the *m*-sound, the next sound she would have covered in the older textbook. But the following week she faced a dilemma: Skip around in the new textbook to continue following the old textbook's sequence and the official curriculum, or follow the new book's sequence? She decided to follow the new book, and went on to its next lesson (*é* and *e*), even though she had already taught these sounds earlier that year. Thus she broke with the official curriculum.

Actual Lesson Structure

As mentioned, the official curriculum laid out a model lesson plan. The left side of Table 3.2 illustrates the sequence of activities suggested for second-grade lessons, which were to take place during four 30-minute sessions over the course of two days. It represents our synthesis of the official curriculum with the official teacher's guide for the older textbook, the two of which aligned closely but not perfectly. Guinean educators explained the lesson sequence as moving from the sentence (step 1 in Table 3.2) to the word (step 2), and from the word to the sound or letter (step 3). For instance, in a second-grade lesson on the *l*-sound, the class would move from a sentence, such as "Moussa, ne lance pas la balle en classe" [Moussa, don't throw the ball in class], to recognition of individual words like *lance* and *balle*. Then they would identify the *l*-sound in *lance, balle,* and many other words. At that point, the class was to move back from the sound to the word (step 4), by blending the new sound to form syllables and words, such as *l* + *an* = *lan* as in *lance.* Finally, they would return to sentences by reading a new text (step 5).

We witnessed few teachers who followed this script exactly. In particular, as an inspector had pointed out, "A lot of colleagues stop at isolation of the letter. Afterward, blending [*la combinatoire*] gets neglected." For example, consider the lesson taught by Madame F in a provincial town in mid-October, illustrated on the right side of Table 3.2. Madame F began the first session of this reading lesson roughly as the Ministry suggested, asking her 72 second-grade students to comment on the picture of the boy Moussa from the book (step 1) and writing the key sentence on the board. Below the sentence, she wrote a string of *l*s, sounding out each as she wrote. She had two boys read the text aloud while they pointed to each word (step 2), then she wrote and explained a new string of words with all the *l*s highlighted in red chalk—*la salade* [lettuce], *des lunettes* [glasses], and so on (step 3).

In a second session the same morning, Madame F began to diverge from the script. For example, she returned to working on words at the board and with homemade flashcards. That afternoon, she returned to word study with the unscrambling of the flashcards. Then she returned to isolation of the sound by having a girl point to all the *l*s on the board, which Madame

Table 3.2 The Official Script and Madame F's Adaptation

Script	Madame F's Lesson on /l/
Day 1 AM session	*Monday AM session 1*
1. Observe picture and produce key sentence	1. Observe picture and produce key sentence
• read key sentence	1, 3. Read key text and note /l/ sound
2. Study words, for example:	2. Identify words (point while reading)
• identify words on board	
• find missing words	3. New words with sound highlighted
• unscramble flashcards	*Monday AM session 2*
	2. Identify words (point while reading)
	3. Point to letter *l*'s on board
	2. Read flashcards and words on board
Day 1 PM session	*Monday PM session*
3. Isolate sound to study:	2. Unscramble flashcards
• isolate letter	2. Identify words (point while reading)
	3. Isolate sound and letter
	?. Dictate to one student.
	1. Read key text.
Day 2 AM session	*Tuesday AM session*
(step 3 continued)	2. Unscramble flashcards
• hunt for words with the same sound	2. Students write words on board
4. Syllable work (blend consonant + vowel)	2. Point to words out of order
	1. Read key text
Day 2 PM session	*Tuesday PM session*
5. New texts or exercises	3. Circle the *l*'s in text on board
	5. Fill in missing letters
	5. Link two phrases to make a sentence

F had highlighted in red. Most unusually, the teacher invited one student to the board and dictated the key sentence for him to write, an activity not in the script at all. The second day of the lesson continued with practice in reading at the board, word study, and some isolation of the sound /l/. At the end of the fifth and final session of this lesson, Madame F introduced some "reading games" (step 5 of the script). In short, Madame F often recycled to prior steps. As the inspector had predicted, she skipped step 4, the blending of sounds—and not for lack of time.

Table 3.3 Suggested Steps Actually Witnessed in First- and Second-Year Classrooms

Suggested Steps	First-Year Classes (N = 3)	Second-Year Classes (N = 9)	Total (N = 12)	Percent of Total
1. Elicit or provide students with the key text	3	8	10	83
• Have students read key text at board or in book	3	7	10	83
2. Study words in the text	2	5	7	58
• Identifying individual words on the board	1	2	3	25
• Identifying missing words	0	2	2	17
• Unscrambling words on labels or slates	1	4	5	42
3. Isolate sound or letter to be studied	2	9	11	92
• Locate sound or letter in a text	1	?	?	?
• Hunt for words with the sound/letter	1	8	9	75
4. Do "syllable work"	0	3	3	25
5. New text or exercises	0	2	2	17

Madame F's improvisation was not unusual. Table 3.3 shows how many teachers in a sample of 12 first- and second-grade classes carried out activities from the Ministry's script during sessions we witnessed.[12] As you see, most of these teachers skipped steps 4 and 5. When we looked at each lesson in more detail, we saw that teachers also "personalized" their teaching in other ways. Some gave special emphasis to a particular activity, such as oral reading at the board, word hunts, or, as in Madame F's case, unscrambling words on flashcards or slates. Other teachers created idiosyncratic classrooms by improvising in less positive ways, moving through sequences of activities in what seemed to us a garbled manner. As a result of the omissions, revisions, repetitions, and occasional garbling of steps in the official script, none of the 16 lessons we observed among these 12 teachers perfectly matched the suggested lesson structure. Not one of these teachers taught a reading lesson in exactly the same way. True, they all drew almost entirely from the same official repertoire of activities, which may explain why outside observers thought Guinean teachers conformed closely to the program. However, their lessons were far from uniform.

Reasons for Improvising

We have seen that the Ministry did not completely control the sequence or pace of lessons or even the choice of textbook. Inside the classroom, we found teachers improvising around the official script, reordering activities and consistently omitting certain steps. Although we had few opportunities to interview teachers away from the inhibiting presence of a director or supervisor, we can hypothesize why teachers exercised this limited autonomy.

Variation in the pace of lessons probably depended on external circumstances such as a delayed start of the school year or interruptions due to teacher illness or training workshops. However, nonconformist choice of the colonial textbook had to be a more personal decision. We suspect that teachers who still used it did so because they were more confident with it, older teachers having taught with it before 1990 and some younger teachers having studied it secretly at home when schools were teaching in national languages (Diallo 1991).

Variation in lesson structure probably resulted in part from recycling certain steps to review or to keep students engaged. However, there were complex historical reasons for skipping step 4, "syllable work" (Anderson-Levitt 2001). Until the mid-1980s, teachers had used the "syllabic" or phonics-based method of the colonial textbook, but the Ministry's new curriculum asked teachers to abandon the syllabic method. Some local inspectors and many teachers interpreted the ban on the syllabic method as a ban on *any* activity involving the blending of sounds. For example, we saw an inspector teasingly warn a teacher that his lesson on blending had been "forbidden" even though it came right out of the pages of the official textbook. We think that inspectors exaggerated their warnings against "syllabic" activities in part because many teachers still found the colonial syllabic method tempting, as we saw above.

At a deeper level, completely scripted teaching was not a true option, for the official scripts had gaps. The sequence of lessons in the new textbook did not match the official curriculum, and there were even small mismatches between the official script in the national curriculum and in the older textbook. Indeed, *any* script must leave gaps (Snyder 1998:227). Thus the Ministry script was anything but "teacher-proof," for to follow it well required a global understanding of the curriculum designers' goals. Ironically, it seemed to us that teachers who followed the Ministry's suggested lesson plan most faithfully included the three or four most competent and confident teachers we observed. They gave the impression that to follow the script carefully and coherently required reflection before and during class.

CONCLUSION

Sometimes, a global reform arrives in a country, local educational decision makers transform it into something new, and then local teachers either creolize it anew or resist it outright. Such was the case, for example, with the entry of outcomes-based education into South Africa (Herbert 2001; Brook Napier in this volume). On first glance, promotion of greater teacher autonomy in Guinea seemed to be the same kind of phenomenon. However, when examined more closely, the pro-autonomy talk that arrived from outside of Guinea turned out not to be a homogeneous global model but rather one side of a transnational debate. The movement toward greater teacher professionalism and freedom is strongly contested in some of its countries of origin.

Inside Guinea, we heard mixed messages about autonomy. Ministry decision makers never adopted the international discourse in favor of broad autonomy. Rather, they appropriated some of its elements to mitigate the difficulties posed by the lack of textbooks and by conflicts between the new textbooks and the official curriculum, but they did so without abandoning the detailed scripts in the national curriculum. Meanwhile, Guinean teachers did not embrace the arrival of new textbooks as a chance to exercise limited autonomy. Rather, for reasons that made sense in terms of their working conditions, they asked for scripted guides. Yet when they actually taught lessons, teachers deviated from prior scripts the Ministry had supplied. Thus Guinean education experienced tensions, if not outright debate, between autonomy and scripting in the Ministry and in the classroom.

There were conflicting messages about teacher autonomy, then, both inside and outside of Guinea. However, this is not to say that tensions within Guinea perfectly mirrored the debates in other countries. Rather, Guineans tended to cast the debate as one between scripted instruction and limited autonomy, whereas the pendulum in the United States and Britain swung between scripted teaching and a much broader ideal of teacher autonomy.

Implications

The tensions and inconsistencies *inside* Guinea do not necessarily challenge the notion of a world culture. World culture theorists acknowledge "rampant inconsistencies and conflicts within world culture" (Meyer et al.

1997:172; see also Finnemore 1996:441). We might acknowledge inconsistent or even conflicting ideas as part of a single, albeit very messy, package of reform. For example, it seems that the Escuela Nueva reform movement and its offshoots in places like Liberia and Namibia may simultaneously encourage teacher decision making and scripted lesson modules (Mantilla 1999; Snyder 1998). Likewise, the reform movement in Thailand gives teachers freedom to design part of the curriculum even as it calls for accountability and hence increased control of teachers (Jungck, this volume).

However, *outside* Guinea, in the United States and Britain, pro-autonomy reforms and pro-scripting reforms are competing movements mounted by different sets of actors. It is not the Holmes Group that advocates scripted instruction, nor are autonomy advocates in England the same people who introduced a semiscripted literacy hour. In the United States and Britain, the pendulum swings between competing models and competing camps of educators. (Indeed, ironically, even as pro-autonomy reform reaches Guinea from the United States, the United States seems to some observers to be "swinging back" toward scripted teaching.) Now, as discussed in the introductory chapter to this volume, George and Louise Spindler have suggested that rather than look for homogeneity in a culture, we look for "cultural dialogues," "expressions of meaning referent to pivotal concerns that [express] *oppositions* as well as agreements" (1990:1, emphasis theirs). We might posit a transnational cultural dialogue—or a transnational cultural *debate*—on school reform. (We noted earlier that the debate is not "global" because some countries like France do not participate in it.) In the present case, educators in Britain and in the United States agree on nothing except the terms of the cultural debate: that it makes sense to struggle over whether teachers get more professional autonomy or less.

However, as the introductory chapter argued, if countries share nothing but a cultural debate—and not all countries at that—there is no reason to expect worldwide convergence in school reform. Moreover, when different camps of actors fight for different models, it may make more sense to interpret the situation as a conflict between competing models rather than a debate within a single culture.

To speak of transnational models of school reform at all, then, we must conceive of them quite differently from the way world culture theorists do. We must imagine reform movements as not only creolized on the ground, but fractured at the source. We must imagine not models but debates, debates that transcend national boundaries (without necessarily touching every country in the world).

Ours is a more conflicted, weaker image than world culture theory's vision of a hegemonic global model of school reform. Nonetheless, we do not mean to imply that fractured, contentious transnational ideas have no significance at all. On the contrary, they inspired the international reformers who brought arguments and resources to bear on Guinea, influencing the shape of the major reforms we described at the beginning of this chapter. Moreover, fractured transnational ideas served as a resource that Ministry experts could appropriate for their own local purposes (again, cf. Rosen in this volume).

The discourse in Guinea about teacher autonomy pointed not to a homogeneous world model, but to actors arguing for local purposes and, in doing so, making use of selected elements of a cultural debate from other countries. If Guinea ever did move from scripted teaching to broad professional autonomy, it might look from the outside like a movement toward a strong if contested ideal from U.S. schools of education. But that is probably not what it would mean inside Guinea. Whether it would represent "progress" in the Guinean context is something the Guineans would have to decide. For the moment, the preference among Guinean leaders seems to be for a much more limited, but reflective, teacher autonomy.

ACKNOWLEDGMENTS

Our research was funded by generous grants from the Spencer Foundation. We thank the Ministère d'Éducation Pré-Universitaire, the Institut National de Recherche et d'Action Pédagogique (INRAP), and many teachers, inspectors, teacher trainers, experts, and international consultants who helped us try to understand their views. Christina Brian, Renee Bumpus, Lara Fawaz, Anna Levitt, Perry Weingrad, and Joelle Younes helped to code the fieldnotes.

NOTES

1. Fundamental Quality and Equity Levels Project.
2. As you will see, many teachers in Guinea hesitated to take up offers of autonomy; however, local resistance to a global movement is not unexpected (Welmond 2002).
3. Sadialiou Barry of INRAP participated in this fieldwork.
4. Ntal Alimasi of the University of Pittsburgh conducted some of the interviews.
5. Success for All and Direct Instruction were the scripted methods recommended by the AFT. As of May 2001, 1800 U.S. schools had adopted Success for All and 300 had adopted Direct Instruction (Northwest Regional Educational Laboratory 2001).
6. We avoid mentioning gender or nationality to better protect anonymity.

7. Increased teacher autonomy might also create more work as trainers for international advisers (cf. Labaree 1992), but our informants showed no sign of such motivation.

8. For example, in the Small Grants School Improvement Project, prefectural officials took a role in purchasing materials for projects rather than disbursing the small grant money directly to schools.

9. Staff of teacher training institutes or educators who had taken an in-service training course of at least 18 months at the Centre de Perfectionnement Linguistique.

10. Teachers who had had in-service training beyond their initial normal school education were more likely to opt for at least limited autonomy. Preference for conformity was strongest (50 percent) among teachers with 10 years or less of teaching experience, but next strongest (40 percent) among teachers with more than 30 years of experience. Being close to retirement in a difficult economy without a pension may have inclined the latter to be cautious.

11. The 14th teacher was teaching a lesson from the second-grade textbook to his combined first- and second-grade class.

12. We left out six first-grade classes observed early in the school year when they were still conducting pre-reading activities. In the remaining 12 classrooms, we did not always observe a lesson in its entirety, but sometimes witnessed sessions from two different lessons, and therefore feel reasonably confident about the data.

REFERENCES

Alexander, R. 2000. *Culture and pedagogy: International comparisons in primary education.* Oxford, UK: Blackwell.

Anderson-Levitt, K. M. 2001. Ambivalences: la transformation d'une rénovation pédagogique dans un contexte post-colonial. *Éducation et Sociétés* 7: 151–168.

Anderson-Levitt, K. M. 2002. *Teaching cultures: Cultural knowledge for teaching first grade in France and the United States.* Cresskill, NJ: Hampton Press.

Anderson-Levitt, K. M., and N.-I. Alimasi 2001. Are pedagogical ideals embraced or imposed? The case of reading instruction in the Republic of Guinea. In *Policy as practice: Toward a comparative sociocultural analysis of educational policy,* edited by M. Sutton and B. Levinson. Norwood, NJ: Ablex.

Anderson-Levitt, K. M., M. Bloch, and A. Maiga Soumaré. 1998. Inside classrooms in Guinea: Girls within the system of interaction. In *Women and education in Sub-Saharan Africa,* edited by M. Bloch, J. Beoku-Betts, and R. Tabachnick. Boulder, CO: Lynne Reinner Publishers.

Baker, V. 1997. Does formalism spell failure? Values and pedagogies in cross-cultural perspective. In *Education and cultural process: Anthropological approaches,* edited by G. D. Spindler. Prospect Heights, IL: Waveland.

Ball, D. L., and D. K. Cohen. 1996. Reform by the book: What is—or might be—the role of curriculum materials in teacher learning and instructional reform? *Educational Researcher* 25 (9): 6–8, 14.

Barr, R., and M. W. Sadow. 1989. Influence of basal programs on fourth-grade reading instruction. *Reading Research Quarterly* 24 (1): 44–71.

Bauman, J. F., and K. M. Heubach. 1996. Do basal readers deskill teachers? A national survey of educators' use and opinions of basals. *Elementary School Journal* 96 (5): 511–526.

Carnegie Forum on Education and the Economy. Task Force on Teaching as a Profession. 1986. *A nation prepared: Teachers for the 21st century.* Washington, D.C.: The Forum.

Datnow, A., and M. Castellano. 2001. Teachers' responses to Success for All: How beliefs, experiences, and adaptations shape implementation. *American Educational Research Journal* 37 (3): 775–799.

Diallo, A. M., K. Camara, J. Schwille, M. Dembélé, and T. H. Bah. 2001. Mobilizing Guinean educators around a primary school quality improvement program. Paper presented at the ADEA Biennial Meeting. Arusha, Tanzania, October 7–11. Available at http://www.adeanet.org/biennial/en_arusha_papers.html

Diallo, M. L. P. 1991. *Enjeux et avatars de l'enseignement du Français en République de Guinée: contexte historique, aspects pédagogique, et perspectives de rénovation.* Doctoral thesis of the University of Bordeaux II, Educational Sciences.

Diané, B., and A. Grandbois. 2000. L'amélioration du système de formation initiale des maîtres par un processus de formation en alternance (institutionnelle et pratique). Paper presented at the annual meeting of the Comparative and International Education Society, San Antonio, March 8–12.

Finnemore, M. 1996. Norms, culture, and world politics: Insights from sociology's institutionalism. *International Organization* 50 (2): 325–347.

Flinn, J. 1992. Transmitting traditional values in new schools: Elementary education of Pulap Atoll. *Anthropology and Education Quarterly* 23 (1): 44–59.

Freeman, D. J., and A. C. Porter. 1983. Do textbooks and tests define a national curriculum in elementary school mathematics? *Elementary School Journal* 83: 501–13.

Fuller, B. 1991. *Growing up modern: The western state builds third-world schools.* New York: Routledge.

Goodnough, A. 2001. Teaching by the book, no asides allowed. *New York Times* A1, 27.

Gursky, D. 1998. What works for reading. *American Teacher* 82: 12–13.

Herbert, R. 2001. Lost in translation: Outcomes based education in South Africa. Paper presented at the annual meeting of the American Anthropological Association, Washington D.C., November 29.

Hoffman, J. V., S. J. McCarthey, B. Elliott, D. L. Bayles, D. P. Price, A. Ferree, and J. A. Abbott. 1998. The literature-based basals in first-grade classrooms: Savior, Satan, or same-old, same-old? *Reading Research Quarterly* 33 (2): 168–197.

Holmes Group. 1986. *Tomorrow's teachers: A report of the Holmes Group.* East Lansing, MI: Holmes Group.

Iredale, R. 1996. Teacher education and international education development. In *Global perspectives on teacher education,* edited by Colin Brock. Wallingford, UK, Triangle Books.

Judge, H. 1992. A cross-national study of teachers. In *Emergent issues in education: Comparative perspectives,* edited by R. F. Arnove, P. G. Altbach, and G. P. Kelly. Albany: SUNY.

Kumar, K. 1990. The meek dictator: The Indian teacher in historical perspective. In *Handbook of educational ideas and practices,* edited by N. Entwistle. New York: Routledge.

Labaree, D. F. 1992. Power, knowledge, and the rationalization of teaching: A genealogy of the movement to professionalize teaching. *Harvard Educational Review* 62 (2): 123–154.

Mantilla, M. E. 2001. Teachers' perceptions of their participation in policy choices: The bottom-up approach of the *Nueva Escuela Unitaria* in Guatemala. In *Policy as practice: Toward a comparative sociocultural analysis of educational policy,* edited by M. Sutton and B. U. Levinson. Westport, CT: Ablex.

Meyer, J. W. 1998. Training and certifying "unqualified" teachers in Namibia: The Instructional Skills Certificate Program. In *Inside reform: Policy and programming considerations in Namibia's basic education reform,* edited by C. W. Snyder Jr. and F. G. G. Voigts. Windhoek, Namibia: Gamsberg Macmillan Publishers.

Meyer, J. W., J. Boli, G. M. Thomas, and F. O. Ramirez. 1997. World-society and the nation-state. *American Journal of Sociology* 103 (1): 144–181.

Meyer, J. W., and F. O. Ramirez. 2000. The world institutionalization of education. In *Discourse formation in comparative education,* edited by J. Schriewer. Frankfurt/New York: Peter Lang.

Northwest Regional Educational Laboratory. 2001. The catalog of school reform models. http://www.nwrel.org/scpd/catalog/index.shtml. [Accessed June 12, 2002.]

Ouyang, H. 2000. One-way ticket: A story of an innovative teacher in mainland China. *Anthropology and Education Quarterly* 31 (4): 397–425.

Paine, L., and L. Ma. 1993. Teachers working together: A dialogue on organizational and cultural perspectives of Chinese teachers. *International Journal of Educational Research* 19: 675–607.

Ramirez, F. O., and J. W. Meyer. 2002. National curricula: World models and national historical legacies. In *Internationalisation: Comparing educational systems and semantics,* edited by M. Caruso and H.-E. Tenorth. Frankfurt: Peter Lang.

Schwille, J., A. Porter, L. Alford, R. Floden, D. Freeman, S. Irwin, and W. Schmidt. 1988. State policy and the control of curriculum decisions. *Educational Policy* 2 (1): 29–59.

Shimahara, N. K., and A. Sakai. 1995. *Learning to teach in two cultures: Japan and the United States.* New York: Garland.

Snyder Jr., C. W. 1998. Glimpse inside structured classrooms. In *Inside reform: Policy and programming considerations in Namibia's basic education reform*, edited by C. W. Snyder Jr. and F. G. G. Voigts. Windhoek, Namibia: Gamsberg Macmillan Publishers.

Spindler, G. D., and L. Spindler. 1990. *The American cultural dialogue and its transmission*. New York: Falmer.

SSP. Service Statistique et Planification. 1997. *Données statistiques, enseignement primaire, année scolaire 1996–1997*. Conakry : Ministère de l'Éducation Nationale et de la Recherche Scientifique.

Stodolsky, S. S. 1989. Is teaching really teaching by the book? In *From Socrates to software. 89th Yearbook of NSSE*, edited by Philip W. Jackson and Sophie Harouttunian-Gordan. Chicago: University of Chicago.

United Kingdom, Department for Education and Skills. 1999. National Curriculum online. http://www.nc.uk.net/home.html. [Accessed July 24, 2002.]

United Kingdom, Department for Education and Skills. 1997–2002. The standards site: The National Literacy Strategy—Framework for teaching YR to Y6. http://www.standards.dfes.gov.uk/literacy/publications. [Accessed July 24, 2002.]

Watson-Gegeo, K. A., and D. W. Gegeo. 1992. Schooling, knowledge, and power: Social transformation in the Solomon Islands. *Anthropology and Education Quarterly* 23 (1): 10–29.

Welmond, M. 2002. Globalization viewed from the periphery: The dynamics of teacher identity in the Republic of Benin. *Comparative Education Review* 46 (1): 37–65.

World Bank. 1994. *World Development Report 1994: World Development Indicators*. New York: Oxford University Press/World Bank.

Zeichner, K., and L. Dahlström. 1999. *Democratic teacher education in Namibia*. Boulder, CO: Westview.

CHAPTER FOUR

GETTING BEYOND THE "ONE BEST SYSTEM"?

Developing Alternative Approaches to Instruction in the United States

Thomas Hatch and Meredith Honig

Amid the growing debates over the globalization of schooling, the United States seems both to embrace and to defy the idea that there can be one model for education. Despite the absence of a national curriculum and despite significant control of education located in school districts, researchers and policymakers in the United States typically lament the limited number of distinct and successful approaches to education. Some have suggested that the striking lack of variation across schools reflects a de facto "one best system" that governs school operations and instruction (Tyack 1974). At the same time, states and districts have pursued several significant and arguably systematic efforts to support the development of alternative educational approaches that reflect the needs and interests of local communities. For example, since the 1970s, the creation of magnet schools has been a popular means of instituting distinctive instructional goals and pedagogies to meet the needs of particular students, employers, and others (Blank, Levine, and Steel 1996). In recent years, the advent of charter schools and small schools reflects a renewed enthusiasm for developing schools that are free from many of the constraints of state and district bureaucracies and more responsive to the concerns of local community members (Bulkley and Fisler 2002; Clinchy 2000; Nathan 1996).

Also, recent systemic and standards-based reform efforts in the United States seek to have it both ways. In many early incarnations, these systemic efforts aimed to give schools and districts the freedom and flexibility to make local decisions about goals and strategies at the same time that they sought greater accountability for results that meet nationally recognized standards (Smith and O'Day 1991). However, these systemic efforts presumed that schools had the capacity to create, implement, and sustain their own approaches: that if policymakers provided aligned policies, national standards, and appropriate incentives, schools would have the conditions, resources, and abilities needed to enable all students to reach high standards of performance (Furhman 1999; O'Day, Goertz, and Floden 1995). Numerous studies suggest that that initial view was far too optimistic; and even after considerable efforts to create a more supportive policy environment, many districts and schools still lack the resources and expertise they need to make improvements (Elmore, Abelmann, and Fuhrman 1996; Goertz, O'Day, and Floden 1996; Massell 1998). As a consequence, states and districts around the United States have created initiatives that seek to build capacity at many levels of the educational system. These include efforts at the state and district levels to institute new forms of professional development, strengthen teacher training programs, develop curriculum materials, and support school planning and data collection efforts (Massell 2000). At the school level, initiatives like the Comprehensive School Reform Demonstration Program (1997) seek to provide schools with "research-based" reform programs that can enable them to develop the capacity that they need.[1]

But the question remains whether these efforts will create more demands and constraints that limit schools' flexibility or whether they will help schools to develop the capacity to pursue their own local approaches (Fuhrman 1999; Hatch 2002; Hess 1999; Spillane 1996). What does it take for schools to develop, implement, and sustain their own successful approaches to instruction? In this chapter, we argue that aligning policies, establishing consistent incentives, and providing resources and technical assistance can help schools to improve, but these efforts are not sufficient. Schools need to develop the capacity to act as autonomous agents that can manage the process of change: They need the means and mechanisms to make improvements in their own instructional approaches without being overwhelmed by the shifting demands and opportunities around them and without having to wait for resources or requirements to come "down" from the district or state.

In order to make this argument, we present case studies of four alternative schools in California—two "progressive" and two "traditional" schools of choice within the public school system.[2] Through these case studies, we suggest that the capacity to develop an alternative approach and maintain a high level of performance over time depends to a large extent on the ability of schools to carry out four organizational practices: monitoring and revisiting their missions, managing turnover, facilitating socialization and professional development of staff, and managing external demands. These organizational practices enable schools to maintain operations and instructional approaches that have been effective in the past while making changes and improvements when necessary. At the same time, these case studies show that these schools, with good reputations and records of success on tests of student performance, have had to work hard to develop their instructional approaches and maintain their levels of performance. Even in a nation concerned with local determination and without a national curriculum, creating alternative schools of any kind is difficult, and developing alternative schools whose operations and instructional approaches depart in significant ways from the schools around them may take more capacity than most schools and communities can manage.

THE CAPACITY TO PRODUCE
ALTERNATIVE INSTRUCTIONAL APPROACHES

All four case studies begin in the 1970s, when groups of concerned parents or teachers sought to create alternative public schools that reflected the educational philosophies that they believed would dramatically improve their children's learning. Both the Emerson School—in a midsize urban district in the Bay Area—and the Dewey School—located in a small, wealthy suburb nearby—were created by parents who were intrigued and inspired by the "open" education movement. The Emerson School has evolved into a K-8 school with multi-age classrooms (kindergarten-second, third-fifth, and sixth-eighth) and a focus on project-based instruction. Currently, with about 300 students, the school serves a diverse population with 43 percent qualifying for free and reduced lunch (compared with a district mean of 54 percent). In terms of student performance, from 1999 to 2001, the students' scores on the Stanford–9 test give it a rating of 7 out of 10 on the Academic Performance Index created by the state, compared with all schools in California, and a rating of 9 compared with schools serving similar populations.

The Dewey School maintains a child-centered curriculum that teachers and parents describe as "supporting the development of the whole child." The school has an enrollment of a little over 400 students who are divided into multi-age classes (kindergarten-first, second-third, fourth-fifth), with students remaining with their teacher for two years. The school has a largely white student body with only 5 percent of students receiving free or reduced-price lunches. The average student scores on the Stanford–9 test are generally in the 80 percent range. In 1999–2001, these scores corresponded with a 10 rating in comparison with all California schools, but the rating in comparison with schools with similar populations ranged from a low of 2 in 2000 to 9 in 2001.

In contrast to Emerson and Dewey, Peninsula and K-8 City employ what are often called more conventional approaches to instruction that include a focus on the development of core academic skills, teacher-led classes, and significant amounts of homework. The K-8 City School, located in the same district as the Emerson School, serves over 500 students, using what staff refer to as an approach focused on "the three 'R's." On average, the school's students consistently outperform many other schools in the district, and, on the state's Academic Performance Index, the school has received rankings of 9 or 10 in comparison with schools with similar populations.

Members of the K-5 Peninsula school describe it as a "structured alternative school" that provides a quiet and orderly learning environment with an emphasis on core curriculum, basic academic skills, and good study habits. The school, in the same district as the Dewey school, has a conventional single-grade structure and serves about 350 students. The school's average test scores are consistently among the highest not only in the district but also in the state.

In order to develop these case studies, we sought to identify organizational characteristics and practices that contributed to a school's ability to manage change, that is, to preserve effective practices and develop new practices in order to maintain high levels of performance over time.

We selected these case study schools based on the strong records of students' performance on standardized tests, a long-standing association with either a "progressive" or "traditional" instructional approach, and recommendations from educators for schools with good reputations within their districts. We collected school documents (such as mission statements, site plans, and materials for prospective students, parents, and teachers) and conducted interviews with the principal, two teachers, two parents, two members of the school district identified by the principal as being particularly knowledgeable about the school, and a former principal or other individual from outside the current school community who could provide some perspective on the school's history. Our interviews focused on three aspects of school operations that previous studies have identified as connected to school capacity: school goals, professional development and relationships among teachers, and parent involvement. The interviews focused on four questions: What were the schools' goals and instructional approaches? To what extent were these maintained or changed over time? How did these schools share information and build knowledge about the goals and instructional approaches among teachers, parents, and the wider school community? To what extent did the schools respond to district initiatives or seek assistance and resources from others?

With their longevity and records of success, these four schools have demonstrated their capacity to develop distinct approaches and to maintain relatively high performance in many of their students. Through the case studies, we describe how four organizational practices—monitoring and occasionally revisiting their missions, carrying out staff development, dealing with turnover, and managing district demands—help these schools to preserve practices that have been effective in the past and to make changes when necessary; and we explore how the relationship between the schools' instructional approaches and the demands and resources in the surrounding environment affect the energy and effort these schools have to expend on these practices. In the process, we show how these schools use these four practices to overcome many of the problems and challenges that often derail reform efforts—including frequent changes in leadership, lack of community among faculty, and crises over inadequate facilities and resources. We also show how they use the practices to deal with the usual changes in personnel and policies in their districts and the shifts in public expectations that provide challenges for many schools.

The case studies also illustrate that the schools use these organizational practices to different ends. At Peninsula, with limited turnover and a strong, veteran staff, they have been able to stick to the same mission and goals without substantial changes. At City, the school managed a significant transition from a veteran faculty used to more conventional teacher-led instruction to a younger faculty more accustomed to the "hands-on" and cooperative learning characteristic of more progressive schools. Both Dewey and Peninsula revisited their mission and established extensive new supports for staff development to overcome significant turnover and extend their pedagogical approaches.

Furthermore, while the members of all of these schools devote considerable time and energy in order to carry out their work, they do not spend their time and energy in the same ways. The members of the schools with more conventional approaches—City and Peninsula—develop their approaches and maintain their performance in relatively informal ways, largely outside of formal meetings, within the demands in the surrounding environment, and without having to seek additional resources. But the staff members of the more progressive schools—Dewey and Emerson—spend extensive time both informally and formally in carrying out these organizational practices. In particular, the members of both progressive schools have engaged in substantial efforts to develop their own formal structures for professional development and significant campaigns to deal with the demands of their districts.

MANAGING WITH WHAT'S GIVEN

With the least turnover among the schools in this study and a qualified, veteran staff, Peninsula has been able to maintain its high record of performance and sustain its "structured" approach to curriculum in relatively informal ways. Even through a period in the 1990s with five different principals in a period of eight years, faculty stayed for long periods of time. As a result, the "Peninsula way" was still carried throughout the school by the veteran teachers, and it continued to stand out as a "traditional" school as other schools in the district embraced the kinds of "hands-on" and cooperative learning and the attention to the "whole child" in use at Dewey. In fact, Peninsula has managed to maintain its "structured" approach to instruction with a conventional weekly staff meeting and by drawing on the professional development offerings in the district and in the surrounding area. When they have needed new staff members, they have usually been able to find experienced, qualified candidates; and although members of the school community have objected to some district policies, school staff have not had to undertake significant organized efforts to deal with those policies or obtain additional resources.

As Janet Stark, a fourth-grade teacher in her second year at the school, explains it, Peninsula is a place where people "help each other and work together," but it's "unofficial." They have a weekly staff meeting to discuss school and district business, and the staff has discussed creating a monthly meeting where teachers from the grade level can discuss their classroom practice, but, for the most part, conversations about their classrooms and their curriculum take place informally. "Like I'll go next door," Stark continued, "and say, 'Oh, I'm going to be doing this. Do you want to do it? This is a project I do.' . . . I think the teachers do that on their own because it's an important part of the way you keep your program going. But at the same time, we haven't had formal meetings."

Similarly, the school has not developed explicit means for new faculty to learn about the school's approach. New teachers at the school can take advantage of trainings that the district offers during the summer and several times during the year for new teachers as well as those changing schools or grades, but there are no formal activities or structures to introduce new teachers to Peninsula's own approach. Instead, during the summer, when teachers are setting up their classrooms for the coming year, veterans like Paula Williams often check in with new teachers on a more informal basis. "We do go around and say, 'Oh, do you need some help?'" Williams explained. "And we just sit around and talk." Those conversations may revolve around "what makes Peninsula Peninsula," materials they need, how to set up their rooms, or other issues.

The fact that most "new" teachers who come to Peninsula are usually experienced teachers from other schools may well reduce the need for a more formal introduction to the school and its instructional approach. "We haven't gotten a lot of brand new teachers," Williams told us. "Like our new first-grade teacher this year, we were talking about 'how are we going to do guided reading?' We're not talking about 'what is guided reading?' We're talking about how you implement it in this particular situation."

When they do need to get new teachers, the school is not simply at the mercy of the district even though the district makes the official hiring decisions. By relying on an informal network of colleagues and contacts, members of the Peninsula faculty spread the word about job openings and often encourage those they feel are compatible to apply. Stark, who had been teaching for almost fifteen years before she came to Peninsula, knew about the school and the job she eventually took because she worked with Paula Williams on a district literacy committee. With what Stark described as a more "curriculum focused" approach than some of the faculty at her previous school, working at Peninsula appealed to her. As a result, when Williams told her, "I'd really love it if you'd come over to Peninsula. There's going to be an opening," Stark thought, "Maybe it's time for a change. . . . So I came over and I looked at the philosophy and read it and talked to the principal and he said, 'We'd really love to have you come over.' So I said, 'Okay, fine.'"[3]

There have been occasions, however, when the school's approach seems to be in direct conflict with the policies of the district. On those occasions, the school has often benefited from the support and influence of parents who have taken it upon themselves to oppose policies that they see as inconsistent with the "structured" approach that they chose for their children. For example, the state approach to mathematics instruction and the district's adoption of a new math program caused extensive concern among both parents and staff in the late 1990s (see Rosen, this volume). As Janet Stark described it, the new program was "conceptual" and did not focus as much attention on the skills and practice of traditional math. In response, parents took it upon themselves to mount a major campaign opposing the district policy and contributed to a statewide movement against what they considered to be approaches to mathematics instruction that were too progressive.

Some school staff have been involved in these organized efforts, but at the school level, the response is often much less formal. Thus, even though a former principal at Peninsula told Janet Stark that the district was "pushing" the teachers—particularly the upper-grade teachers—to use the new math program, she found that the district gives teachers the freedom and flexibility to use what they prefer. "And so you could get a recommendation that you really should be doing this [the new math program]. But if the students are doing well, and they're succeeding, no one's going to say, 'Well no. You have to do this.' It's never that emphatic." As a result, even though the district went ahead with the math program, to a large extent, Peninsula's parents got what they wanted. Many of the teachers continue to use the same math materials they've used all along, while some, like Stark and some of the newer teachers at the school, choose to use some aspects of the new program along with some more traditional exercises and materials that have been in use at the school for some time.

Although the school has been able to deal with periodic conflicts with the district and has maintained its instructional approach without developing many new structures or practices, changing times and the gradual incorporation of new faculty over the years have taken a toll. As a result, in the 2000–2001 school year, Peninsula began an examination of its guiding philosophy and core beliefs. One goal of that examination was to look at what distinguishes Peninsula from other schools in the district, since, as Williams, the second-grade teacher explained, "everybody's beginning to look more like us," or as Stark, the fourth-grade teacher, put it, "most of the regular neighborhood schools are shifting over to a more traditional, structured approach." Although many have welcomed the process, the effort has not been without controversy; in particular, in a reflection of the continuing tensions between the school and parents, a number of parents complained that David Summers handpicked the community representatives on the committees. Nonetheless, the committees produced a new statement of core beliefs in the spring of 2001 and Summers expects to turn shortly to the second step of the process, when the school considers how to put those values into practice.

CHANGING WITH THE TIMES:
INFORMAL AVENUES TO SOCIALIZATION
AND INSTRUCTIONAL IMPROVEMENT

For many years, City School, like Peninsula, maintained its "three 'R's" approach with little turnover and relatively little attention given to schoolwide meetings or professional development. When Julianne Fredericksen came to the school in 1996 as the new principal, she was well aware of City's strong reputation, but she was also aware of concerns at the district level about some racial tensions and the responsiveness of the school to the needs of all students. Many of the teachers "were outstanding teachers for children that learned a certain way," she explained. "But they weren't outstanding teachers for children who learned in ways that were not real traditional. . . . our African American students and those with different learning styles, their needs were not always met and addressed." She suggested that the school's approach needed to change with the times and become more responsive to the changing population and needs of its students.

In order to manage this change, however, the school relied largely on informal means, without creating formal schoolwide structures or activities. Instead, coupled with a buyout offered by the district, Fredericksen saw class-size reduction in the mid-1990s as an opportunity to bring in new teachers with a somewhat different approach. As she put it: "When we took on a couple of new teachers with class-size reduction, that's really when we first started to change." In the process, she focused on developing a faculty that was, as she put it, "more user friendly" and that included teachers who were not only good teachers but teachers who "could work with all children in all types of families." In particular, she looked for teachers who were not only certified to work with English Language Development (ELD) students, but who were also trained or had experience meeting the needs of diverse students. And she encouraged veteran teachers as well to get certified to work with ELD students so that the school would not have to have separate classes for ELD students and general education students. "I wanted every teacher to get their certification for a couple of reasons," Fredericksen explained. "One, it allowed me the most flexibility at placing students. And two, then I didn't have to deal with the parents who felt that their child was in the class that was [designated] ELD, that they were in a class that was inferior. . . . This teacher here who had only General Ed children, her test scores for sure were going to look better than this teacher's who had General Ed and English Language Development children." In order to determine whether prospective teachers met their criteria, Fredericksen and her assistant principal carefully screened all the candidates to look for those who "fit" the new directions for the school. "My assistant principal was African American," Fredericksen explained, "and we asked the questions like 'how will you teach African American students that don't tend to respond to lectures? . . . And how will you work with their families?' And if we didn't like the answers to those questions at an initial interview, the teacher did not move on."

Beyond managing the turnover of a veteran staff, the school invested some time and effort in professional development, but on a relatively informal basis. Rather than using existing meetings or creating explicit structures for professional development, Fredericksen concentrated on creating supports for the new teachers she hired and establishing an atmosphere or climate that would encourage everyone to stretch their skills. She "protected" the new teachers and really made them feel welcome, as James Anderson, a fifth-grade teacher who has been at the school for over 25 years, explained. "She had meetings for the new teachers. Got them special materials. Anything that they put on a list, she seemed to get for them." In order to build on the work of teachers' colleagues, Fredericksen also encouraged teachers to visit one another's classrooms or the classrooms of mentors or model teachers in other schools.

Although many veteran faculty initially objected to the changes at City, over time, as those veterans retired and new teachers were hired, complaints subsided, and Fredericksen, as well as the teachers and parents we spoke with, feels that the school has a renewed sense of creativity, spirit, and vitality. As James Anderson put it, "we've finally left the Flintstone era." But concerns about whether or not the school is losing its focus on the three 'R's remain. Although Anderson sees the changes as leading to "more critical and deeper-thinking students," other veterans—and some parents—worry that the hands-on approaches that many of the new teachers have brought are watering down the standards of the school. Megan Lawrence, who joined the school a year after Fredericksen became principal, told us: "When I came in, [the veteran teachers] would complain, 'Oh, look at those grades, those sixth-grade teachers, you know, all those kids are getting As and Bs. That's going to bring down the quality of the school.'" Lucy Simmons, a parent, who has had children throughout Fredericksen's tenure, has been happy with the changes. "I feel like [my children] are still learning. They're still getting good facts," she reported. But she stated the concerns of some parents clearly: "Are they [the students] still learning the same things? Are they still learning as much as they used to? And I think some parents feel like, no, they're not learning as much as they used to."

CREATING COLLECTIVE PRACTICE:
FORMAL AND INFORMAL EFFORTS TO
STRUCTURE ORGANIZATIONAL LEARNING

Although the members of Dewey and Emerson continue to refer to the philosophies upon which they were founded, school staff have substantially expanded and refined their instructional approaches and spend much more of their time and effort working together in developing that approach than they ever did before. Even in an environment that ostensibly allows and encourages the development of alternative approaches to schooling, they have had to go far beyond conventional resources and structures to develop and refine their instructional approaches and maintain their success. At both schools, those efforts have involved extensive reexaminations of their founding philosophies and missions, and those reexaminations have, in turn, led the members of both schools to establish a wide range of formal and informal efforts to discuss their curricula and classroom practice, find and socialize new teachers, and deal with district demands.

Revisiting Missions

At Emerson, they embarked on what Diane Kirsch, a former lead teacher, termed "the goals thing" in 1989. Given threats of closure from the district, significant teacher turnover, and a changing student population, the faculty was concerned about how to improve their curriculum and ensure (and demonstrate) that their students were making adequate progress. As a result, the school devoted a series of meetings over several years to a process in which faculty examined the founding tenets of the school and considered which ones remained powerful and relevant. "And then we looked at where did we want to be in ten years, what were our goals for ten years. And then using that as a base we started looking at different reform efforts," Kirsch explained. Ultimately, in the mid-1990s, that process of exploration led them to become a part of Project 2061—a national initiative spearheaded by the work of the American Association for the Advancement of Science to develop a variety of instructional models to improve instruction in the sciences. Through their participation in that effort, they developed the approach to project-based instruction that has become a cornerstone of their work and a distinguishing feature of their school.

At Dewey, when Charlene Moore, the current principal, came to the school, the mission "was written everywhere," she told us. "But nobody gave it to anyone to read, so the [new faculty] didn't know." In response, Moore established a committee to rearticulate the "Dewey way." She selected equal numbers of faculty and parents to serve on the committee and asked a parent who worked at a nearby high-tech company to head the committee and act as facilitator so that Moore could be simply an "equal participant." The committee examined everything that had been written about the school, and Moore shared what she had learned from her conversations. They also looked at the history of the school by inviting the founding principal to talk about the school's origins. They used all of these materials, Moore explained, "to see if we [could] come up with some easy way of saying what Dewey is all about." All of this took almost two years, and they emerged with a short brochure, a concise description of the school, and a renewed sense of a shared mission.

Managing Professional Development and School Culture

Although the reexaminations at Dewey and Emerson all began with a focus on their missions and goals, the rearticulation of missions that were put on websites and in brochures was not nearly as important as the concrete ideas and structures it generated in helping to foster continued collaboration. At Dewey, the group that worked on the brochure continues to meet to discuss school goals and activities: "The group thought the task was done when we did the brochure. Oh, no . . . ," Moore said, shaking her head. Instead, she turned the group into a "community relations committee" that meets on an ad hoc basis. Moore sees their role as "the gadflies for the Dewey philosophy." Two other groups also meet to help Moore address issues related to the "culture of the school." One group meets weekly on Friday afternoons, and she refers to them as "the historians, the old guard": "When they come to me and say, 'Well, we always did that at Dewey,' they have to tell me why," Moore explained. She sees their charge as keeping the history alive so that when they leave, "the history remains." Moore also meets with the leaders of the PTA and the school site council twice a month (before their own meetings) to discuss new ideas and recent developments. All three groups serve as sounding boards for her ideas and help her to distribute the responsibility for reflecting on and maintaining the school's culture.

Similarly, at Emerson, adoption of the project-based approach in the course of the school's reexamination inspired the development of a variety of organizational structures and practices. In particular, the staff has established an extensive set of meetings that includes: a meeting of the whole staff every Monday; weekly meetings of the faculty at each developmental level; and then biweekly committee meetings that are often used for planning purposes, including a "professional development team meeting" and a "lead team meeting." The whole-staff meetings alternate between a focus on professional development and a focus on business. For the developmental team meetings, participants have specified five different topics they address in order "to make sure that those things that we're thinking about at the whole-staff level get actually put into practice and evaluated," explained Claire

Marx, a fourth-grade teacher. These topics include curriculum planning and development; sharing results of schoolwide inquiries in which teachers are engaged; looking at student work; ways to address the needs of students experiencing particular difficulty; and standards and standardization across classrooms and projects.

Beyond the regularly scheduled staff meetings, faculty at both schools have opportunities outside of meetings to see and hear what their colleagues are doing. At Dewey, these include informal conversations in the lunch room and elsewhere in which faculty discuss their curricula and the progress of their projects, and what grade 4–5 teacher Georgeanne Kim referred to as "the sharings": occasions when faculty invite their colleagues and often other students to see the results of particular projects. At Emerson, although the intense meeting schedule and a lack of a common lunch period make it difficult for many of the teachers to get together informally, the teachers also get the chance to observe the results of one another's work all the time. First, because they share students during the projects, they can see the progress that students make with different teachers. Second, during their meetings and retreats, they spend time as a whole staff looking at student work throughout the school. Third, faculty at Dewey have created a formal structure—a semiannual open house for the community—in which students share the results of their projects. Through this mechanism (and the free flow of people from one room to another), student learning is made visible for members of the community as well as for the staff. Occasions when the staff has identified inconsistencies have led to schoolwide initiatives like the focus on writing, in which the staff spent several years developing rubrics and jointly scoring student work.

Dealing with Turnover

Whenever possible, both schools also take advantage of mentoring or support programs that are available to new teachers (such as California's Beginning Teacher Support Program), but a lot of work with new teachers at Dewey and Emerson goes on informally as well. At Dewey, for example, veterans like Warren Lovejoy, a grade 4–5 teacher, have occasionally had formal, paid responsibilities for working with new teachers, but in recent years he and others like Georgeanne Kim simply find times when they can "check in" with new teachers. As Marielle Henkel described her experience in her first year: "I'm right next door to Georgeanne Kim, and we have a door that connects our classrooms and that door was open a lot the first year. And I really feel that Georgeanne went out of her way to make things really work for me." At Emerson, the retreats, the regular staff meetings, sharing students, and the open houses all enable veteran teachers to see what new teachers are doing and vice versa.

The high levels of turnover experienced by both schools have contributed to the need for such extensive efforts to socialize new teachers and to promote collective work on curriculum and instructions. In fact, although Fredericksen at City could use turnover to make changes, at Dewey, Moore suggested that class-size reduction was "almost the death of Dewey," because it required the hiring of so many new teachers who were relatively unfamiliar with the "Dewey way." At Emerson, although they maintained a small stable staff throughout much of the 1970s and 1980s, the aging of the staff, the demands of class-size reduction, and the growth in students assigned to the school by the district have all led to a steady influx of new teachers.

As a result, both schools have also attempted to influence the hiring process, encouraging applicants whose interests and qualifications seem to match the school and weeding out those whose do not. Although each district actively works to recruit new teachers, it cannot rely on those efforts, because as one Dewey parent involved in the hiring process put it, "the district had no clue" what the school was about. As a result, both schools work to develop informal networks to build knowledge of their approaches among applicants and find applicants with the skills to carry those approaches out. Dewey, for example, has benefited from the "word of mouth" among members of teacher education programs in the area who direct some of their student teachers to seek out placements at Dewey because it stands out among schools in the area that take a "progressive" approach. At Emerson, in order to increase their chances of hiring the "right" candidates, the staff has always spent considerable time reaching out through their colleagues in the district to try to spread the word about the school, and, over the years, the school has made a particular (though not always successful) effort to hire teachers of color. They also met with many of the eligible candidates in the district to try to talk them out of applying. "We'd say, there's all of this work," Kirsch told us. "This is what we do. We know that teaching is a hard job anywhere, but the expectations here are really over and above what they are at other places, and it's not the right place for everybody." To ensure that no one misses this point, they also provide candidates with materials about the school, including their extensive meeting schedule, and ask candidates to sign a "commitment sheet" that spells out what they are agreeing to. Through these efforts, staff can make more informed choices about whether or not applicants will fit at the school, and prospective teachers arrive already tapped into the school's mission and practices.

Managing District Demands

In addition to having to develop their own means of supplementing their districts' offerings for staff development and their hiring procedures, Dewey and Emerson also have had to devote significant staff time and energy in a wide range of formal efforts to deal with the actions (and often inaction) of their districts. In particular, at Emerson, the facilities problems and threats of closure from the district meant that staff and parents had to work closely together to mount extensive campaigns to lobby district staff and school board members to preserve the school. They also had to fight for resources: When Diane Kirsch served as lead teacher, she pored over the financial records of the school and found that at a time when the per pupil expenditures for schools in the district with a similar population of students were about $4000, Emerson was getting only $2500 per student from the district. As a result, the members of Emerson have had to expend substantial time and energy seeking additional funding, resources, and technical assistance from a significant number of other support providers including Project 2061, the Bay Area Coalition for Equitable Schools, and the Bay Area School Reform Collaborative (which provide technical assistance and resources for particular approaches to school improvement).

The staff at Dewey never had to deal with the kinds of threats that Emerson did, and like the other two schools in the study, did not have to seek substantial funding from other sources (although all three benefited from parent-led efforts to raise additional funding). Nonetheless, even in an affluent district, with what members of both schools described as a generally supportive relationship with the district, the staff at Dewey has had to work to get what they want. In those efforts, parents have been an integral part of efforts to negotiate with the district, but as at Emerson, for the most part, the impetus and organization has come from the school. For example, while all other schools in the district use a standardized report card, Moore called the superintendent to explain that Dewey does not use report cards: "We tell the parents, 'We don't compare children with each other,'" Moore described the conversation, "'and that's what a report card does—it's a comparison of you with somebody else.' He said, 'Well, because we are looking at standards, it is not a comparison of the child.'" After what Moore called a "long philosophical discussion," the superintendent told her to request a waiver and he would consider the issue. But, as Moore put it, "I was not going to do a report card, I'd made up my mind." In response, she organized a day-long study session in which several parents, teachers, a board member, and a representative of the district reviewed Dewey's approach to assessment. In

the wake of this demonstration, the board member reported back to the
school board that Dewey has a reason not to do the report cards and needed
an exemption.

In some cases, the conflicts between Dewey and Emerson and the demands of their districts may simply reflect a difference in style or approach or
the reality that districts often have to make difficult choices about how to al-
locate scarce resources and deal with a wide variety of different schools and
constituencies. But they also reflect the fact that the districts themselves experience considerable turnover and face a considerable workload that makes
it difficult to learn about the individual approaches of alternative schools like
Dewey and Emerson. As a result, staff at Dewey and Emerson also have to
take the time to help district staff to understand what they need. For example, when a new superintendent arrived, Moore found that she had to spend
time every year negotiating with him and explaining why the school needed
to use the professional development days devoted to the district's agenda to
hold its own retreats. In order to deal with this recurring problem, after the
school reexamined its philosophy, Moore worked proactively with the parents
and teachers on her community relations committee to build an understanding of the school among the superintendent, board members, and district
staff. To do so, they held a series of dinners at which they conducted the same
kind of orientation they provide for new parents to the school: describing its
history, mission, goals, and practices. They also invited the superintendent
and board members to visit the school and talk with the teachers "anytime,"
which many of them did. From Moore's perspective, the results justified their
investment of time and energy. "Things are better now. . . . I felt for a while
like I wasn't the principal, I was this public relations person, and it was kind
of tiring, to tell you the truth, but all worth it." Worth it because the following year the superintendent allowed Dewey to use several days dedicated to
district-based professional development for its own purposes.

Despite the sometimes conflicting demands and lack of district support
experienced by these schools, both Dewey and Emerson have managed to
maintain their own distinctive approaches. At the same time, their experiences so far suggest that nothing has been determined for certain. If the past
is any indication, they will continue to experience issues with faculty
turnover and will continue to have to deal with policies and common practices that conflict with their missions. In particular, in addition to developing their abilities to provide an effective educational approach to students,
the members of both schools will have to continue to develop their relationships with an ever-changing group of parents, community members,
and district officials in order to ensure that they have the resources and flex-

ibility they need. In the process, they will have to continue to debate and discuss the merits of their approach and demonstrate their success in ways that are compelling to others.

DISCUSSION AND IMPLICATIONS

Taken together, these activities—establishing and revisiting missions, creating structures to support staff development and collective practice, dealing with turnover, and managing external demands—provide a glimpse of what it takes for schools in the United States to mount their "own" instructional approaches. Like the strings of a guitar, the schools can use these functions to tune their performance.[4] The passage of time, changes in the supply of teachers, and shifts in policies and public expectations can all result in schools becoming out of tune with their surroundings or can put pressures on schools that contribute to mission drift or break down common understandings or relationships among members of the school community. Under these circumstances, tightening the mission—rearticulating it and ensuring that the members of the school community understand it—can make it clearer to job candidates and staff members whether or not their ideas fit with those of the school, provide a basis for making decisions about how and when to manage external demands, and enable schools to reestablish collective work. Or a school like City can change practices to be more in step with the times by focusing time and energy on hiring and developing new staff who may then help to establish a new mission. Similarly, investing more time and energy in professional development and in supporting the exchange of ideas among staff may minimize the need to focus specifically on rearticulating the mission.

At the same time, the schools in this study demonstrate there is no simple formula or approach that can ensure that a school can develop or maintain its own approach. Simply "having" the features of these successful schools—having a mission, developing a plan for recruitment and hiring, establishing an approach to professional development, and mounting efforts to manage external demands—is not sufficient. The members of these schools have to know when to devote time and energy to these different functions and how much to expend on them when they do, and ultimately they have to be able to find the time and energy to follow through.

Given the experiences of the members of these schools, which have unusual records of success and resources and flexibility that may not be available to many other schools, the challenges of managing the process of change and maintaining a locally developed approach to education seem particularly daunting. However, when successful, managing the mission, turnover, professional development, and external demands creates a culture that sets expectations for the type and quality of instruction and for the nature and extent of interactions with colleagues. At Emerson, that culture helps to sustain the school's approach and enable newcomers to carry it out. As Claire Marx describes it, "seeing people who work here teach and being colleagues with them just set a standard, and really influenced how I learned about being a good teacher." At City, focusing on informal support for newer teachers as well as managing turnover helped to bring new ideas and a new peer culture that made some veteran teachers uncomfortable but that encouraged others to extend their own development.

The pressure that staff feel from being a part of these schools may also increase the chances that individual teachers can resist district policies and practices that do not fit with those of the school. At Peninsula, for example, Stark may feel more pressure *not* to use the new math program from the veterans and parents she sees every day than she feels pressure to use it from the district officials, whom she sees less frequently. Similarly, at Emerson, the extensive commitment to internal staff development, sharing of students, and collective examination of student work dictates classroom instruction much more directly than occasional interactions with district staff.

All in all, these findings suggest a fundamental rethinking of the capacity that schools need in order to be successful. Although capacity is often treated as the amount of resources a school needs to accomplish a particular goal or perform a particular function—what Cohen and Ball (1999) call a "space and storage definition"—in the view presented here, the ability of a school to establish and maintain a successful instructional approach depends upon the relationship between the school and the surrounding environment. Thus, the capacity of a school is affected both by its ability to receive and implement certain policies and practices and by its ability to act on and manage the surrounding environment. If a supply of qualified teachers is not readily available, then the members of a school—like those in this study— have to be prepared to go out and find them. If policies and expectations in the district are not consistent with those of the school, then the school staff have to be able to go out and change them.

Correspondingly, although developing any alternative approach may take considerable capacity, the more distinctive a school's approach is, the more time, resources, and energy the school may have to invest in developing and maintaining it. Thus, at Dewey and Emerson, with their progressive philosophies, staff have to carve out time and expend their energy in developing alternative arrangements, carrying out their own professional development, and mounting campaigns to deal with their districts. City and Peninsula, with their more traditional approaches, can take advantage of the predominant structures and available resources and put their time and energy to work directly on curriculum and instruction. Under these circumstances, even within a system that encourages the development of alternative approaches, it may be the far easier of the two paths for schools—particularly schools with few resources and without records of success—to adopt approaches that are more consistent with the dominant system. Even in a country where it often appears that the preferred philosophy is to let a thousand flowers bloom, most of those flowers may end up looking the same. But, perhaps efforts to reconceptualize and reexamine what it takes for schools to develop and maintain alternative approaches will enable district and school personnel to negotiate and reconcile the policies that come from the top down with the goals and practices that come from the bottom up.

ACKNOWLEDGEMENTS

Work on the case studies described in this chapter benefited from the support of the William and Flora Hewlett Foundation and the Carnegie Foundation for the Advancement of Teaching; however, the responsibility for the conclusions expressed lies with the authors. The authors also wish to thank Karen Herbert, who participated in the data collection and analysis for the case studies; Connie Behe, who assisted in the preparation of the manuscript; and Kathryn Anderson-Levitt for her comments on previous versions of the chapter.

NOTES

1. Congress established the Comprehensive Reform Demonstration Program in 1997 in order to "help raise student achievement by assisting public schools across the country to implement effective, comprehensive school reforms that are based upon scientifically based research and effective practices" (http://www.ed.gov/offices/OESE/compreform/2pager.html). In order to implement these reforms, Congress appropriated $145 million in FY 1998, $260 million in FY 2001, and $310 million in FY 2002 to provide qualifying schools with up to $50,000 a year for three years.

2. In these "schools of choice," instead of district assignment to the school based on geography or other factors, admission is based on a lottery among students from the district whose parents have applied to the school.

3. In addition to this informal agreement between the previous principal and Stark, the principal still had to go through a formal process in which the

hiring was discussed with the district and other principals, with the final determination made by the district.

4. See McDonald (1996) for a related use of the idea of "tuning" in school reform.

REFERENCES

Blank, R. K., R. E. Levine, and L. Steel. 1996. After 15 years: Magnet schools in urban education. In *Who chooses? Who loses?* edited by B. Fuller and R. F. Elmore with G. Orfield. New York: Teachers College Press.

Bulkley, K., and J. Fisler. 2002. *A decade of charter schools: From theory to practice* (CPRE Policy Brief No. RB-35). Philadelphia: University of Pennsylvania, Consortium for Policy Research in Education.

Clinchy, E. 2000. *Creating new schools: How small schools are changing American education.* New York: Teachers College Press.

Cohen, D. K., and D. L. Ball. 1999. *Instruction, capacity, and improvement: CPRE research report series.* Philadelphia: University of Pennsylvania, Consortium for Policy Research in Education.

Comprehensive School Reform Demonstration Program. 1997. http://www.ed.gov/offices/OESE/compreform/2pager.html

Elmore, R. F., C. Abelmann, and S. H. Fuhrman. 1996. The new accountability in state education policy. In *Holding schools accountable: Performance-based reform in education,* edited by H. Ladd. Washington, D.C.: The Brookings Institution.

Fuhrman, S. H. 1999. *The new accountability* (CPRE Policy Brief No. RB-27). Philadelphia: University of Pennsylvania, Consortium for Policy Research in Education.

Fuhrman, S. H., and R. F. Elmore. 1990. Understanding local control in the wake of state education reform. *Educational Evaluation and Policy Analysis* 12 (1): 82–96.

Goertz, M. E., R. E. Floden, and J. O'Day. 1996. *The bumpy road to education reform* (CPRE Policy Brief No. RB-20). New Brunswick, NJ: Rutgers University, Consortium for Policy Research in Education.

Hatch, T. 2002. When improvement programs collide. *Phi Delta Kappan* 83 (8): 626–639.

Hess, F. M. 1999. *Spinning wheels: The politics of urban school reform.* Washington, D.C.: Brookings Institution.

Massell, D. 1998. *State strategies for building local capacity: Addressing the needs of standards-based reform* (CPRE Policy Briefs No. RB-25). Philadelphia: University of Pennsylvania, Consortium for Policy Research in Education.

Massell, D. 2000. *The district role in building capacity: Four strategies.* (CPRE Policy Brief No. RB-32). Philadelphia: University of Pennsylvania, Consortium for Policy Research in Education.

McDonald, J. P. 1996. *Redesigning School.* San Francisco: Jossey-Bass.

Nathan, J. 1996. *Charter schools: Creating hope and opportunity for American education.* San Francisco: Jossey-Bass.

O'Day, J., M. E. Goertz, and R. E. Floden, 1995. *Building capacity for education reform* (CPRE Policy Brief No. RB-20). New Brunswick, NJ: Rutgers University, Consortium for Policy Research in Education.

Smith, M. S., and J. A. O'Day. 1991. Systemic school reform. In *The politics of curriculum and testing,* edited by S. Fuhrman and B. Malen. Bristol, PA: Falmer Press.

Spillane, J. P. 1996. School districts matter: Local educational authorities and state instructional policy. *Educational Policy 10*(1):63–87.

Tyack, D. 1974. *The one best system: A history of American urban education.* Cambridge: Harvard University Press.

CHAPTER FIVE

RESISTANCE TO THE COMMUNICATIVE METHOD OF LANGUAGE INSTRUCTION WITHIN A PROGRESSIVE CHINESE UNIVERSITY

Huhua Ouyang

This study argues that at the grassroots level of specific school communities, individual students and teachers always have their own interpretations about and appropriation of any curriculum reform launched by the state from above. The case rests on my longitudinal participant observation of how and why students and teachers in a pro-reform university in mainland China complained about native English-speaking teachers from the West.

The historical context against which this case is set is the great national reform of the past 20 years. In moving toward openness to the outside world and modernization, China needs personnel that are more independent, creative, and productive than it has had. However, the traditional methods practiced in the Chinese classroom, which are based on conservative ideologies and rote learning, have been inadequate to fulfill this goal. Therefore, since the late 1970s the education authorities have carried out a national campaign of reforming the curriculum toward a more Western-style liberal pedagogy. Communicative Language Teaching (CLT), which advocates student centeredness, communicative learning, a humanistic approach, and practical learning, has become increasingly prevalent. "CLT" has therefore become a buzzword, especially in modern English-language teaching in China. In fact, CLT has gradually become popular globally since the late 1970s, spreading its main tenets that language is better acquired by learners as free, equal, independent, and rational decision makers using the language in authentic communicative contexts, in a process of discovery. This is in sharp contrast to the traditional notion that language is a set of rules that should be taught and learned by rote in the form of knowledge transmitted from the authoritative teacher to passive learners. Believing that CLT can bring about learner autonomy and creativity, the Chinese English-language teaching authority has established its reform in line with this international trend (Dzau 1990).

To speed up this reform process, hundreds of native speakers of English from countries such as the United Kingdom, the United States, Australia, and Canada have been invited to work in China. These native speakers of English were invited as foreign teaching experts to demonstrate "advanced" CLT.[1] They are often assigned to teach in programs jointly managed by the Chinese education authority and international agencies such as the British Council or the British Overseas Development Agency. These foreign experts have usually been granted important responsibilities, including teacher training and in-service teacher development, textbook composition, and test design, to help spread the CLT type of reform nationwide, from the central cities to the more remote regions (Hayhoe 1989). Although no official documents explicitly say so, it is generally assumed that the foreign teachers are experts or authoritative role models for the new and "advanced" methodologies like CLT, since they originate from the West, as do the reforms. In fact, they are often literally addressed as "foreign experts," a term used interchangeably with "foreign teachers." Their expert status, relative to their Chinese colleagues, is also evident in preferential treatment in their living conditions and welfare. They exercise the privilege of making independent decisions in their work and of not having to conform to the collective and uniform decision making of the institutions to which they are assigned.

Given this situation, it is logical to assume that the foreign experts and their teaching will be and should be accepted by the Chinese wholeheartedly. Nevertheless, although many of the foreign experts have had very successful experiences working in China, it is an undisputable fact that many of them have not. To cite one of these foreign experts, the situation has been such that "a large number of foreign teachers returned from China with dampened enthusiasm, feelings of disappointment and in some cases bitterness and rancor. . . . Their Chinese hosts often privately feel that these foreigners are a weird lot and wonder if it is worth all the time, energy and money they expend on having them" (Maley 1990:103).

As compellingly proven by Martin Schoenhals' (1993) ethnography of a middle school in Beijing, students' evaluations can be very useful in revealing the real state of teaching. However, relatively little substantial work has been available concerning students criticizing their foreign teachers. Interestingly, the scant literature on such criticism springs not from the students themselves, but rather from foreign experts reflecting on problems they encountered with their students (for example, Oatey 1990). Interpretation or explanation of the students' criticism backed up with the insider's knowledge is also rare (but see Dzau 1990a; Cortazzi and Jin 1996a, 1996b). I believe that by critically examining the details of how Chinese students and teacher colleagues evaluate the foreign experts, and why, we can come to understand the extent to which the top-down CLT reform is appropriated, creolized, or resisted at the level of the school community.

Although this paper examines criticisms of the English experts, most Chinese students still regard having a native English-speaking teacher as a luxury and many foreign teachers do have a very successful teaching experience in China. It is not the intent of this study, nor is it even possible, to measure the merits of the foreign experts against their demerits in the Chinese students' eyes. Rather, one of the chapter's purposes is to point out that often the same students who have praised the foreign experts file complaints about them.

The present study is largely ethnography based. The voices of the Chinese students and teachers as well as of the foreign teachers will be presented as verbatim as possible, however limited or even possibly "biased" the informants' perspectives could be. Besides collecting and presenting the data in a naturalistic or qualitative way, I have provided some of the background or historical information necessary for explaining the opinions. My explanation revolves mainly around the research site as a community of practice with more or less shared norms, "forms of membership, and construction of identities" (Lave and Wenger 1991:123). I have close contact with the research site. After four years of undergraduate study at the site as an English major, I have worked there as an English teacher since 1983. I have served as a teacher educator and a course coordinator for two teacher education programs jointly organized by the Chinese Ministry of Education and the British Council for over ten years, and currently serve as the head of the English Department at the site. Presumably, my knowledge about this community would grant me the status of a reliable insider, or a longitudinal participant observer in a loose sense.

GUANGWAI AND ITS 20-YEAR JOURNEY OF CLT REFORM

The research site Guangdong University of Foreign Studies (hereafter Guangwai) is located in the city of Guangzhou (also known as Canton), which, with a geographic closeness to Hong Kong, enjoys many more opportunities than most Chinese inland cities to have close interaction with the outside. The interaction is especially enhanced by millions of Cantonese-speaking Chinese overseas who are the relatives of the Guangzhou citizens. Guangwai is the city's main institution for producing qualified personnel to communicate in nine foreign languages, including English.

Since the late 1970s, the British Council and some North American agencies have played an important role in assisting Guangwai to become the most pro-reform institution in Chinese tertiary education. The English Department, which is the research setting of this case study, has received much foreign aid as well as support from the education ministry to implement various reform-related projects. Dozens of its Chinese teachers have been sent to pursue a master's in applied linguistics or English-language teaching with British Overseas Development Agency scholarships for one year in some of the top universities in the United Kingdom.

CLT had been introduced into the Department of English in Guangwai in the late 1970s, largely due to an influential textbook series called *Communicative English for Chinese Learners* (better known as *CECL;* see Li 1987) for use by first- and second-year English majors. This pioneer project is regarded as "an attempt at a thorough adoption of the communicative approach" (Dzau 1990b:7) and a very successful one (Malcolm and Malcolm 1988). *CECL* replaced the old textbook for undergraduate English majors beginning with the class of 1979–1981. I still remember its shocking impact as an experience totally different from what we had been used to in a classroom dominated by the traditional methods. The traditional method, which combines grammar-translation and audiolingual methods, requires students to read a short passage of classical literature (300–400 words) intensively, take notes on the teacher's lectures on the language points and grammar rules involved in the text, memorize the list of its vocabulary and phrases, then practice the grammar in exercises. These activities usually took two weeks of eight class hours each. With *CECL,* we were given about 50 pages of texts and exercises to cover in the same amount of time. The texts were authentic English newspapers, magazines, or daily communication. We skimmed, scanned, and listened to dozens of passages while guessing the meanings of the main ideas and new words from the context. Meanwhile, we were to participate in dozens of oral English activities of simulation and role-play. The overriding criterion was that insofar as our meaning got conveyed, our communication would be valued as successful— regardless of how broken or half-baked our linguistic form might be. The teacher's role was supposed to be that of friend, monitor, and mentor.

In evaluations at the end of the second year, half of the students praised *CECL* highly, saying that it had aroused their interest to learn, provided opportunities to use English, and increased their confidence. The other half, usually the older ones, criticized *CECL* sharply, complaining that it wasted their time and that they learned nothing solid. The older students got support from the teaching staff members who were relatively senior in age and more conservative and who were in charge of the third and fourth years of undergraduate study. Together they arrived at the decision, over the complaints of the other half of the students, that from the third year on, all students should revert to grammar study and intensive reading of literature classics.

From the mid-1980s, *CECL* took root and gained predominance at Guangwai. These CLT-formulated textbooks spread to the first two years of all departments with English majors. Gradually the teachers of *CECL* have become those young graduates who got master's degrees in applied linguistics, where CLT represented progress. Other teachers more senior in age and in position, who were more critical of *CECL*, have retreated from the first two years' teaching groups to teach more traditional courses (such as literature, intensive reading of classics, writing, grammar, and translation) to the third- and fourth-year students. In fact, some of the most senior teachers left the Department of English for good and founded their own department, partially due to this split in opinions over *CECL*. On the other hand, the textbooks have also gone through revisions since 1985: Some lists of vocabulary, Chinese translations, and more grammar exercises were added, responding to complaints from those teachers who were educated in traditional methods or were skilled in using them. Although there are still divergent opinions about *CECL* and CLT among Guangwai faculty members, today as a collective they tend to see *CECL* as a source of institutional pride vis-à-vis colleagues from other universities. Moreover, Guangwai students have become critical of the "traditionalism" of secondary teacher trainees who come to the institute (Ouyang 2000b).

Beginning in the mid-1980s, as part of the endeavor to prove that CLT was good and *CECL* effective, Guangwai went on to spend over ten years successfully reforming the national university entrance examination toward a more communicative competency-oriented test in place of the former grammar rules–dominated one. Guangwai has set up the first program of linguistics and applied linguistics both at the MA and PhD levels in China, and the program is now the only key national institution of linguistics and applied linguistics.

Given this background, it would be interesting to see why Guangwai students and teachers, given their identity as pro-CLT reformers, should have made complaints about its foreign teachers.

DATA AND INFORMANTS

The English Department, with a working faculty of about 70 active instructors, has hosted on average over a dozen foreign experts each year. The
data for this study were collected from 1997 to 2000 as part of a larger project (Ouyang 2000a). The research questions asked in interviews, questionnaires, and focus group discussions were: 1) What do Chinese students most
often complain about regarding their foreign teachers and their teaching? 2)
How do students make such complaints? 3) How do the leadership and administration handle the complaints? 4) How do foreign teachers involved
take the complaints? and 5) Are there any other factors the informants see as
relevant to the complaint issue? This chapter focuses on the first question.

Six *xuexi weiyuan,* or "class committee members in charge of academic
affairs," each representing one fourth-year class of about 30 students (hence
representing altogether 180 students), were chosen to give a detailed report
on the research questions. One of these students' responsibilities was to collect their classes' opinions on teachers' teaching twice a year. So far, these students had had nine courses taught by foreign teachers, including oral
English, reading-writing, American society, American literature, organizational behavior, and British literature. All these students had had *CECL* in
their first two years of study in Guangwai.

Of the faculty colleagues whom I interviewed, nine were finally selected
to be my key informants. Among them, there were department heads, coordinators, and teachers with highly varied experience in working with foreign
colleagues. Given their duties and background, they had been most heavily
exposed to foreign teachers and had been closely involved in and had functioned crucially in many complaint events. I also consulted two Guangwai
Foreign Affairs Office officials in charge of the evaluation, who had many
routine interactions with the foreign experts. Seven foreign experts gathered
in a focus group discussion with me and shared their opinions and experiences about students' complaints. The data from the students and foreign
experts were in English, and other data in Chinese (sometimes interspersed
with English) were subsequently translated by me.

Because of my research questions, I elicited many reports of complaints.
Nevertheless, to quote one student representative:

> Before I write down my complaints about their teaching, I would like to
> make it clear that these complaints do not apply to all of them. I should
> admit that some of them are excellent teachers. They are skillful in making
> class lively and interesting, and we do learn something from their teaching.

However, compared with their Chinese colleagues, foreign experts were criticized more often by their students. My own estimate, based on my observation as course coordinator and on my foreign experts' focus group discussion, was that about 60 percent of the foreign experts drew criticism about their teaching, while about 10 percent of them received very serious criticism (such as when students as a class wrote a written report to the dean's office demanding replacement of the foreign teacher involved). About 10 to 15 percent of foreign experts were highly praised for their teaching, with half of that number having actually "gone native," that is, having learned about the Chinese traditional methods of teaching or somehow met the local requirements for excellence (see a case below for details).

GUANGWAI COMPLAINTS ABOUT FOREIGN TEACHERS

As we shall see, much of Guangwai students' criticism about their foreign teachers and their teaching have to do with things that are not exactly related to differences between CLT and traditional methods. Rather, they are often results of students adopting the standard for ideal teachers and teaching and using it to measure the foreign teachers' teaching performance and roles. However, it is exactly my point that any grand methodology or approach is substantiated by and contextualized in such students' expectations and ideals as a major part of the preexisting practices in any school community. In addition, we can use this reality to check how the new methods and its agents fare at the grassroots.

"Just Improvising":
Complaints about the Teaching Method

The most frequent complaint, agreed upon by over 70 percent of the Guangwai students, teachers, and leaders, was the lack of systematic organization and linearity in the foreign teachers' classes, which resulted in a lack of a sense of achievement for the students. In almost all courses taught by the Chinese teachers in Guangwai as well as other universities in China, some fixed textbook is used. For this uniform textbook, every step of the lesson is given in the teachers' book and is usually followed, and a good sense of linearity is usually not a problem with students. Even in the *CECL* course, all the activities were detailed by the teachers' handbook. Teachers teaching *CECL* had spent hours collectively preparing the lessons each week, planning which parts to emphasize or skip over, the correct answers to the exercises, the pace of the lessons, and other issues, so as to "standardize" the

process of teaching and the teaching methods. The amount of lesson preparation for teachers is so meticulous that one foreign expert who used to teach
at Guangwai exclaimed, "It is amazing how they manage it, and it would be
extremely difficult to teach in that way for foreign teachers" (Tim Boswood,
formerly the British Council senior lecturer, personal communication). As
the working language used in the collective lesson planning and preparation
is Chinese, foreign experts do not usually participate in it. This abstention
is often taken as a privilege enjoyed by the foreign experts.

In other courses using more traditional methods (where the student representatives were in their fourth year), teaching in general was very often equated
with accurate delivery or transmission of prescriptive knowledge from the
teacher as an expert to the student as an apprentice. Students in these courses
learn not in the form of active participation, as in questioning and performing
activities where they could use or explore the language in class, but by taking
notes and reviewing the lecture after class. After consulting various dictionaries
or grammar manuals for almost every new word and expression in the text to
be covered in the lesson, teachers transfer all those language points onto the
blackboard—each with several concrete examples for illustration of its usage in
varied contexts—for the students to copy into their thick notebooks. Most examinations still stress how well the lessons as discrete or countable pieces of
knowledge have been learned by students, who prepare themselves for the examinations by reviewing and digesting their notes. Therefore, the quality and
quantity of the notes students can get from a class is a major indicator of their
capacity to learn, as well as an important reassurance of their sense of achievement and progress in learning. A department head remarked:

> The most common complaints from students is that foreign experts like
> to talk wild in their teaching, from the south of the earth to the north of
> the sky; they improvise too much, and this makes it difficult for students
> to prepare for their teaching. This is especially frustrating for those good
> students; we all know that they want to take detailed notes from the
> lessons. Without that, they felt they had not got anything useful.

A student representative confirmed:

> We are not very used to their style of teaching. Most of them only speak,
> speak and speak, instead of writing something on the blackboard. Chinese
> students, incapable of taking notes, put down nothing when a marvelous
> speech is over. Moreover, I find that many foreign teachers' presentation is
> not as orderly or systematic as their Chinese counterparts'. I think it is also
> an obstacle for students to take notes.

"Kindergarten Teaching":
Complaints about the Activity Type

The "fun" style of teaching, which is typical of foreign teachers, is to encourage students to move around in the classroom physically through various game-like activities. The teachers would use personal anecdotes or jokes to make the teaching lively, and invite free and active participation from the students. However, these teaching methods were not well received by the Chinese students, who often thought that these foreign teachers did not teach "seriously." Students did not just complain about the oral classes but also about writing, literature, and culture-and-society lessons. One typical complaint from a student representative was:

> The foreign teachers usually treat us like kindergarten kids, making us sing, dance, and interact like children in oral classes while we are wondering what on earth is happening.

It is my observation that having fun whilst learning is an idea alien to most Chinese teachers and students, who believe that learning should be hard and that achievement is proportional to the hardship endured. Chinese teachers will praise those who study hard but not those who are born intelligent—a view supported by scholars studying Chinese psychology (for example, Gow et al. 1996). Almost all primary school teachers remind their first-year pupils that learning in kindergarten can and should be fun but that the fun is over forever when they move to primary school. In other words, for most mainland Chinese children from the age of six or seven, learning is socially constructed as a serious and tough task. Since such fun activities are typical of learning only in kindergarten in the Chinese education system, they often gave rise to a feeling among students that they were being humiliated.

"They Don't Correct Our Mistakes":
Complaints about Error Correction Methods

Most of the foreign teachers I interviewed criticized the traditional and still prevalent Chinese practices of teachers correcting mistakes extensively and of not allowing peer or self-correction. Some of them argued that errors are not only tolerable but also valuable and normal in the learning process. Despite the fact that all students in Guangwai have learned English for two years already with *CECL,* with its strong emphasis on fluency over accuracy in oral English communication, when it comes to writing in English, almost without exception they insist that correction of their mistakes in the English language is an indispensable means for them to improve their learning. In addition, the amount of time that the teacher spends on correction is seen as a clear indication of the teacher's teaching quality and attitude. The exclusive power of teachers as the standard for evaluating students and students' expectation that teachers will correct them explain the common practice of denying students free access to teachers' books or the answer key; such information falls strictly within the teachers' sovereignty. It was not until very recently that some market-oriented bookstores began to sell such materials to students, which is creating tremendous pressure and a threat to teachers' authority (Zhang Xiaoling, personal communication). Even students in Hong Kong, according to a recent study (Li and Chan 1999), still tend to favor error correction from their teachers.

Thus the foreign teachers' practice of no error correction led to complaints about foreign teachers being lazy and not fulfilling their responsibilities. Two students reported:

> They don't correct our mistakes. I know I have made some grammar and lexical errors, and how disappointing to see that they are not corrected with the right answers. How can I learn anything from this teacher if she does not do her job?
>
> The foreign teachers like to give us praises as remarks to our written work, such as "excellent," "very interesting," and "fascinating"; it seems to us that all they look for is enjoyment for themselves as readers. They don't know that we don't need that; what we need is the provision of the correction that all Chinese teachers painstakingly do.

These remarks are confirmed by my own teaching experiences from 2000 to 2002 in an academic writing course, where my efforts at exploring peer review and peer feedback as a substitute for the teacher's marking encountered persistent resistance and failure.

"Self-Made Unsuitable Material":
Complaints about the Syllabus

In China, teaching has almost always been centered on or prescribed by one single standard textbook. To some extent, this is comparable to the role of the Bible for Christians or manuals for novice car drivers. A unified textbook has social implications. Firstly, teaching has to be conducted as a collective rather than individual action. Secondly, decision making is not the business of ordinary teachers but of top central authorities. Thirdly, Chinese textbooks have to go through a strict censorship to exclude potentially politically incorrect content (Dzau 1990a). Lastly, standardization of a textbook makes standardized testing and centralized evaluation easy. It is also desired that, by using the standard textbook, those teachers in the less developed areas can have a more guaranteed teaching quality.

However, coming from a much more individualist and liberal system, these foreign teachers all displayed a radically different opinion about the use and role of textbooks in learning and teaching, which too was a source of potential complaint, as shown in the following excerpt from a student:

> Our teachers like to use materials composed from only they know where, in bits and pieces, and we get worried about what can be learned from such materials since most of them don't seem to have a coherent or consistent theme or subject. In fact, very often they are just what the foreign teachers are interested in themselves.

The Chinese teachers and leaders confirmed the use of idiosyncratic criteria in selecting teaching materials as biased and lacking quality control. One of the departmental leaders said:

> The foreign teachers usually complain that the textbooks we use are too much and meaningless for them to teach with, and they don't listen to our idea that they could use partly what is in the standard textbook and integrate into it something they think good. So they chose to have their own. But after a while of taking some extracts from here and there, they end up with using no materials at all! They use their family anecdotes, personal experiences, and anything that happens to be their interest.

To the Chinese students, who are used to the reassuring standard textbooks, the way foreign teachers use materials of their own choice could lead to a sense of insecurity and confusion. While the self- or tailor-made materials could, as confirmed by most of the focus discussion group, better cater to students' interests or proficiency levels, the Chinese teachers and students at the interviews claimed that such matters as "family anecdotes, personal experiences, and anything that happens to be their interest" should be kept for chats between friends outside of class and should not be used in class. Studies of Chinese social psychology seem to support this view in that in-class teaching is regarded as public and formal communication and thus should engage as much as possible in the real business of learning about the syllabus content (for example, Bond 1991; Gow et al. 1996).

"Biased": Complaints about the Grading Criteria

Complaint about the criteria used by some foreign teachers to evaluate students' performance, although it involved fewer students than the prior five complaints, was the one most strongly expressed by the Guangwai informants. This is understandable, since all students tolerate least well any low grades, from foreign and Chinese teachers alike, especially when the students think they had good reason to believe the criteria used were unfair, as in this case. As exemplified in a student's words:

> In writing courses, some foreign teachers regarded highly those works they think [of] as "creative," "with individual opinions," or "interesting" according to their own views, and those who usually achieve high marks because of their good language and structure in other teachers' classes were labeled as "lack of opinions," or "not critical enough." So the top students in our class got not as good scores as those middle-level ones with "opinions" that foreign teachers liked.

When I confronted the foreign teachers in the focus group with this complaint, most of them did not deny it. Instead they stressed that they had the right to award marks according to what they thought to be the right and fair standard. Actually, some of them articulated that it was an individual teacher's business or authority to make whatever judgment concerning his or her teaching, once they have been offered the contract, which to them implied authorization or endorsement for such decision making. During our focus group discussion, they were somewhat shocked when I informed them that in New China, in spite of the fact that classrooms have four walls, teachers have in practice no authority in the final say concerning students' performance. The teacher's assessment of students' performance in their class is always subject to a superior judgment. Such higher-order judgment could come from several sources: political instructors (who are responsible for students' daily life matters ranging from political participation in campaigns, to social and collective activities, to dormitory relationships with other roommates, and even to matters of personal hygiene), head-teachers, course coordinators, heads of department, and student representatives; all of them could intervene in the assessment of individual cases. Should there be any dispute regarding their assessment, the superior authority always has the discretion and right to alter an assessment, often to the detriment of the teacher who made the original assessment. The foreign teachers in the focus group discussion had no idea of how their Chinese colleagues managed to align their grading closely with other colleagues teaching the same course. In fact, Chinese faculty members often discuss in groups or consult each other to ascertain what criteria should be used in their evaluation, a routine job conducted in the collective lesson-preparing time (Ross 1993).

"Not Caring": Complaints about Interaction with Students and about Foreign Teachers as Role Models

It seldom failed to surprise most foreign teacher informants that they were judged not only by their teaching performance inside the classroom, but also by their way of interacting with students outside of class, not only as professionals but also as citizens in terms of their virtue as a moral exemplar, judged by the Chinese standard. For instance, in Guangwai, Chinese teachers usually are authoritarian figures in class when lecturing, showing a serious face and focusing only on academic matters. However, they usually balance this with a much more humanistic mothering attitude after class when interacting with students in private settings. (This observation of mine is supported by many studies of Chinese leadership such as Bond 1991; Bond and Hwang 1986.) A model teacher said, "I will never give away the authority to decide what is best for them in class, in syllabus choice, in teaching and learning methods." Yet it is the good teacher's responsibility to approach students to find out what help they need in private visits and contacts, for instance in students' dormitories. Through such informal and private interaction, teachers could establish close and personal relationships and knowledge about students, and offer help accordingly. The same teacher went on to say, "I am very much like my students, listening to them as friends after class, and trying to boost their independence and self-esteem."

Many Chinese students expressed their displeasure at not seeing this in-class and out-of-class balanced teacher-student interaction pattern in their foreign teachers' behavior—for example, in responding to students' questions after class. One student articulated:

> It is shocking for some of us to find out that in fact our foreign teachers don't like us to ask them questions during the class break time or after class—they indicate that we take advantage of their private time—"exploiting them," as one of them said. This is especially disappointing if you contrast it with the kind of warm and encouraging faces they show us during their teaching time; some of us even suspect that they are putting on an act in class to be so egalitarian and friendly while they withdraw into such cold selfish reaction when we approach them outside the class.

It is a well-established finding (Cortazzi and Jin 1996a, 1996b) that in general Chinese students do not raise questions actively during class time, for they are concerned that both the teachers and they themselves might lose face. Cortazzi and Jin reported that if teachers fail to answer the question, they would very likely lose credibility and authority in public, which will in turn make the student lose face for making this happen—something improper in the eyes of the majority of students (1996b). However, as indicated above, such a pattern of question and answer is compensated for by what happens during class breaks, when students actively approach teachers for individual questions. I have learned in my years of practice as a teacher that in fact most Chinese teachers embrace those questions, seeing them as positive indicators of students' respect for their authority and face, as well as a high motivation to learn in class. As a result, the break time very often becomes a time of more intensive labor for many teachers. Often the foreign experts' enthusiastic invitation of active participation from students during the lesson would create a very strong impression on students (and perfectly logically) that: "If you are already so friendly in class, you must be doubly friendly after class."

Almost all foreign teacher informants considered the practice of asking questions during the break as taking advantage of their private time and their friendliness. As some foreign experts said in the focus group discussion: "I need the rest time to clear my mind"; "I had no rest at all"; "It is such a large class and they don't know how tired you can be." With these feelings, many foreign teachers said they were either very reluctant or simply refused to be bothered during the break as well as in their private time. They preferred to give fixed office hours, which students are not used to, for about 15 Chinese teachers use one office. Another concern of the foreign teachers about mixing with students outside of class time, one shaped by their own sociocultural interpretation, was a fear that such behavior could be suspected as sexual abuse on the part of the foreign teachers. Actually, since all the teachers and students live on campus, visiting students in their dormitories can be convenient and is conducted under public surveillance, hence there is no chance for teachers to abuse their authority through such visits.

Most of the foreign teacher informants ridiculed the notion when I informed them that they were also judged on the basis of how they comported themselves outside of class, in their private time. Their habit of going to bars to make casual friends, dating local girls,[2] wearing casual clothing, sitting on desks in class, and so forth, could possibly convey a negative impression to students, who hold strict standards for their teachers' proper behavior as citizens.

"Self-Conceited":
Complaints about the Teachers' Attitude

The ways of teaching exemplified above were often seen as evidence of a self-conceited attitude. Most comments could be deemed as a negative interpretation of the kind of autonomy and individuality fostered in foreign teachers' teaching, as in the following quote from a department head:

> We would not blame them too much for their use of unsuitable materials or teaching in their favorite ways; they are foreigners after all and they have no knowledge about Chinese learners. But they don't actively ask advice of their students or Chinese colleagues about students' needs and expectations for English teaching in the Chinese context. Some of them are stubborn and insist on their own justification and their autonomy in teaching their ways even when they are criticized and suggested/asked to change their ways.

This critique may sound biased and unfair, since almost all the foreign teachers in the group discussion and interviews claimed that they had made an effort to inquire about what students needed, something they were proud of doing as part of their democratic tradition. However, it was *how* they asked that made the difference: The Chinese ask in private, and ask the class committee members. I observed in my ten years of work as a course coordinator for foreign teachers that since they failed to ask in this local, "appropriate" way, they usually either elicited some very superficial answers from those "who talk but don't know" rather than those "who know but don't talk," or complained about the passivity of students when asked to give them answers they wanted (see Ouyang 2000a for a fuller description).

When the complaints became serious and students explicitly demanded a replacement for the foreign teacher involved, the Guangwai department leaders, course coordinators, or the head teachers of the class would talk to the foreign teacher and suggest or even demand change. Yet, most such warnings or orders were made so indirectly, out of concern for preserving the foreign teacher's face and dignity, that most of the message was lost on the teacher. Sometimes the foreign teacher was not sensitive to the seriousness of the matter. In other cases, the foreign teacher would argue back strongly, demanding hard evidence of students' complaints and justification. Only about one third of them would accept the complaints and try to patch things up with changed action in class (Ouyang 2000a).

One major reason for this defensive reaction to the complaints was probably the self-perception of being a foreign expert. According to the Chinese

coordinators and students, not a few foreign teachers seemed to be indeed mis-
led by or carried away with the title of "foreign expert," believing that their
way of teaching English was precisely the much more advanced know-how
that they were paid to demonstrate and pass on to the Chinese students (cf.
Harvey 1990; Maley 1990). Quite a few foreign teachers responded to Chinese suggestions with an attitude explicitly verbalized in the group discussion
by some: "How could you expect no change when what you wanted from me
in the contract was something so different from your routine?" One intervie-
wee went to the extreme of saying, "Isn't it stupid to criticize me for what I am
paid to demonstrate to them? I am the best that can happen to them!"

Guangwai's Ideal:
A Foreign Teacher Who Has "Gone Native"

It is best to conclude this complaint section by presenting a piece of Guangwai's evaluative comments on a foreign teacher who won the 1994 "Friendship Award" offered to 100 foreign experts working in China (from
Guangwai Application for 1994's Friendship Award, in-house document).
This teacher later took part in the Beijing celebration of the 45th Anniver-
sary of the People's Republic, which included a grand dinner hosted by the
premier. Despite the fact that such foreign teachers "gone native" are few in
number—five or six in about 20 years of Guangwai's history—they have
demonstrated that when given long enough employment and enough sensi-
tivity and effort, foreign teachers can become as good as their best Chinese
counterparts. In fact, many aspects of this foreign expert's behavior can be
seen as more traditional and "Chinese" than most Chinese teachers' behavior. The comments succinctly reveal that in Guangwai the excellence of a
teacher consists of what is best of both the newly Western-imported and the
traditional expectations for ideal teachers. Here are parts of the comments:

> . . . He has very good professional ethics, is responsible, holds a sincere at-
> titude, and works with conscientiousness. . . . Many Chinese teachers like
> to discuss with him about teaching, for he is very friendly, agreeable, and
> always ready to help. He listens to the evaluation results with modesty, and
> adjusts his teaching in a timely fashion. . . . He adjusted the teaching ma-
> terials to students' level. He prepares lessons seriously. He motivates stu-
> dents well. He lectures with good linearity, clarity, and emphasis. Students
> can follow him, comprehend him, learn a lot by heart. Besides, he strictly
> requires students to do everything. He treats tests and assignments very se-
> riously. He gives assignments and very seriously marks them, and makes
> detailed feedback. . . . He treats students with heartfelt warmth and gives

patient tutoring. In his private time, he actively approaches those students who are backward in study, and discusses their problems, gives objective suggestions, and helps them overcome difficulties in learning. He actively marks students' work in addition to the assignments, he helps them to revise their drafts, tutors them in oral tests, and actively approaches students during break time in class and after class to talk with them, which not only enhances their relationships as teacher and students and enriches his understanding about students, but also provides students with plenty of opportunities to practice oral English . . .

CONCLUSION

Complaints arise mostly out of disappointment at things and people that one has too high an expectation for or an unrealistic assumption about. From this study, it seems logical to conclude that Guangwai contradicted itself in that the foreign teachers were invited to demonstrate what Western pedagogy was like and yet were punished for doing exactly that. This self-contradiction is shocking if one considers Guangwai's status in Chinese English-language teaching reform as one of the most radical promoters in Chinese tertiary education, where CLT reform has been implemented for over 20 years. If Guangwai could react to the reform like this, other universities or institutions in less developed places, being less pro-CLT, would most probably resist the reform and the reformers further and more deeply. On the other hand, it seems equally logical to claim that since the reform is a Chinese one and the setting is in China, the foreign teachers should fulfill the expectations of the Chinese if they are the employers or clients—to do as the Romans do in Rome. After all, Guangwai has had its own 20-year history of CLT implementation.

Perhaps the most important fact to bear in mind is that obviously both sides use their own versions of how CLT should be conducted, and yet neither had ever realized exactly what sociopolitical ramifications the foreign teachers' CLT practices in the Guangwai community would bring about for individual teachers and students there. It was a real case of experiential learning through discovery for both sides. Both had to improvise because policies from the top are always abstract and provide only general guidelines for direction and resource allocation. When the general concepts land on the ground of specific schools, they are inevitably substantiated and mediated by individual students and teachers (including the foreign teachers), with specific histories and idiosyncratic ways of seeing and doing things, in a particular institutionalized mechanism of social control and interaction like

Guangwai. In other words, there would be hardly parallel readiness and synchronized adjustment for innovation at top and bottom societal structures. Indeed, as I have shown elsewhere (Ouyang 2000a), the social organization of Guangwai—job and residence immobility, egalitarian social welfare, and antibureaucratizing administration—supports and demands many features of the traditional system.

Thus the CLT umbrella terms have become creolized in Guangwai, taking on characteristics of the Guangwai community of practices. We have seen that the process of Guangwai's appropriation of CLT was never smooth, simple, or clear-cut. Rather it was always full of inconsistencies, struggles, and a dualistic psychology of "not only . . . but also." In a sense, it was the failure or inadequacy in appreciating and practicing/appropriating this CLT with Guangwai characteristics that pushed foreign teachers to the margins, even though they had been invited by the Chinese Ministry of Education as presumably authoritative role models for the "Western-imported advanced technology."

We see that in Guangwai the faculty and students as a collective were proud of its *CECL* and CLT pioneer status in China, for these had brought them benefits such as free study in the United Kingdom and the acquisition of Guangwai-style fluent English and appropriate communication skills. However, the conflicts of interests resulting from splits in values toward CLT had driven some of the faculty staff to retreat elsewhere. We see that the extreme CLT practitioners coexisted with colleagues skilled in traditional methods, although each occupied different grade levels as its territory. The same inconsistency led Guangwai to criticize rural middle school teachers sticking to traditional methods as being too conservative and backward while complaining that the foreign teachers from the West were too liberal and self-conceited. We also see that, living in a ritualized communal routine of collective lesson planning and preparation, Chinese teachers had taken great pains to standardize the choice of textbooks, teaching pace, and grading criteria. In this context, they saw the foreign experts' lecturing style, class activities, methods of error correction, teaching materials, and marking as individualistic and egocentric. Students habituated to ritualistic and collective teaching thus found foreign teachers' lessons missing the familiar sense of linearity, achievement, and psychological security.

Finally, we see that in the transition from traditional methods toward CLT, the best teachers of Guangwai, including the few foreign teachers gone native, have married the best features of the Western liberal and humanistic approach and of Chinese traditional practices, with the latter's emphasis on moral shaping. It is these teachers who can maximally satisfy students' "greedy" demand for a teacher: These teachers not only allow and encourage them to be free, egalitarian, and self-assertive, but also set up a strict moral role model and paternalistic leadership for them while offering authoritative delivery of grammar and vocabulary knowledge of English. Failure to observe this "not only CLT but also traditional best" standard caused many foreign teachers to be seen as not fair, not caring, or not morally sound enough.

All of these observations suggest that at the grassroots of schools, there is no such thing as a universally identical CLT or global schooling, as is often assumed by people who believe in or make top-down policy. CLT at the level of global schooling remains largely something general and vague. Bottom-up forces, such as the community's history and institutionalized practices and the stakeholders' past experiences, interpretation framework, and subjective agency can determine the exact extent to which CLT is accepted, rejected, or creolized. The locals always appropriate the top-down schooling, giving it meaning on their own terms, for, after all, it is their business, their classroom, and their well-being in their own territory that is at the stake.

NOTES

1. "Foreign" is a term used by the Chinese people with no intent of discrimination.
2. In the West these behaviors are usually hidden as part of a teacher's "private" life, for if they were known, they could also have a detrimental effect on a teacher's reputation, explained one foreign teacher.

REFERENCES

Bond, M. H. 1991. *Beyond the Chinese face: Insights from psychology.* Hong Kong: Oxford University Press.

Bond, M. H., and K. K. Hwang. 1986. The social psychology of Chinese people. In *The psychology of the Chinese people,* edited by M. H. Bond. Oxford: Oxford University Press.

Cortazzi, M., and L. Jin. 1996a. English teaching and learning in China. *Language Teaching* 29: 61–80.

———. 1996b. Cultures of learning: Language classrooms in China. In *Society and the language classroom,* edited by H. Coleman. Cambridge: Cambridge University Press.

Dzau, Y. F., ed. 1990. *English in China.* Hong Kong: API Press.

Dzau, Y. F. 1990a. Teachers, students and administrators. In *English in China,* edited by Y. F. Dzau. Hong Kong: API Press.

———. 1990b. How English is taught in tertiary educational institutions. In *English in China,* edited by Y. F. Dzau. Hong Kong: API Press.

Gow, L., J. Balla, D. Kember, and K. T. Hau. 1996. The learning approaches of Chinese people: A function of socialization processes and the context of learning? In

The psychology of the Chinese people, edited by M. H. Bond. Oxford: Oxford University Press.

Harvey, P. 1990. A lesson to be learned: Chinese approach to language Learning. In *English in China,* edited by Y. F. Dzau. Hong Kong: API Press.

Hayhoe, R. 1989. *China's universities and the Open Door.* Armonk, NY: M. E. Sharpe.

Lave, J., and E. Wenger. 1991. *Situated learning: Legitimate peripheral participation.* Cambridge: Cambridge University Press.

Li, D. C. S., and A. Y. W. Chan. 1999. Helping teachers correct structural and lexical English errors. *Hong Kong Journal of Applied Linguistics* 4 (1): 79–102.

Li, X. J. 1987. *Communicative English for Chinese learners* (series of textbooks). Shanghai: Shanghai Foreign Language Education Press.

———. 1990. In defense of the communicative approach. In *English in China,* edited by Y. F. Dzau. Hong Kong: API Press.

Malcolm, I., and K. Malcolm. 1988. Communicative language teaching materials for China: The CECL project. Paper presented at the Regional Seminar on Materials for Language Learning and Teaching: New Trends and Developments. Regional Language Center, Singapore, April 11–15.

Maley, A. 1990. XANADU—"A miracle of rare device": The teaching of English in China. In *English in China,* edited by Y. F. Dzau. Hong Kong: API Press. [Originally published in 1984, *Language Learning and Communication: A Journal of Applied Linguistics in Chinese and English,* (out of print) NY: John Wiley.]

Oatey, H. 1990. Teacher training in the PRC: Influence of sociocultural factors. In *English in China,* edited by Y. F. Dzau. Hong Kong: API Press.

Ouyang, H. H. 2000a. Remaking of face and community of practices: An ethnographic study of what ELT reform means to local and expatriate teachers in today's China. Unpublished PhD dissertation, City University of Hong Kong.

———. 2000b. One way ticket: A story of an innovative teacher in mainland China. *Anthropology and Education Quarterly* 31 (4): 397–425.

Ross, H. A. 1993. *China learns English: Language and social change in the People's Republic.* New Haven: Yale University Press.

Schoenhals, M. 1993. *The paradox of power in a People's Republic of China middle school.* New York: M. E. Sharpe.

CHAPTER SIX

WORLD-CULTURAL AND ANTHROPOLOGICAL INTERPRETATIONS OF "CHOICE PROGRAMMING" IN TANZANIA

Amy Stambach

This chapter examines the culturally contingent roles of parents in the organization and operation of public schools (called "government schools") in Tanzania. Specifically, it explores a recently introduced English-language primary school program, proposed and initiated by U.S. Protestant missionaries working in Tanzania, and several Tanzanian parents' reactions to this program.

Until recently, Tanzanian parents had been viewed by policymakers and school administrators as largely irrelevant to the administration of government schools. With the exception of a handful of *wazazi,* or "parent-run" schools, parents had had little say in their children's schooling. For roughly the past two decades, however, Tanzanian parents have played increasingly central roles in the administration and organization of public education, and the role of administrators has waned in the face of growing involvement by parents and nongovernmental organizations.

The apparent waxing and waning of the roles of parents and the state, respectively, occurs in the context of a more general and seemingly worldwide debate over the organization and administration of mass education. Is mass education best administered through public agencies, or is it better controlled through market forces that revolve around ideas of supply/demand and competition? Should public officials oversee the organization and quality of formal schooling, or is this task better handled by local communities and parents? Such questions have generated discussions about whether, and if so how, parents ought to be provided with a "choice in education." One premise of discussions about school reorganization is that "choice" will maximize competition and ensure efficiency. In the United States, educational choices include charter school programs, voucher programs, home-schooling options, and open enrollment programs in which students attend public schools outside their residential attendance areas (see Good and Braden 2000). In Tanzania, educational choices include open enrollment programs and English-language programs such as the one described in this chapter.

Many studies have focused on choice programming in the United Kingdom and the United States (e.g., Ball, Bowe, and Gewirtz 1995; David 1993; Good and Braden 2000), and several have discussed how choice programs complement processes of school administrative decentralization (for example, Berman 1999; Bray 1999), but few have explored the specific trajectories that move choice programming around the world, and few have examined the interpretive frameworks within which people conceive of and develop programs and practices they call "choices in education." My focus in this chapter is on differences between Tanzanian parents' and U.S. missionaries' views about choice programming in education. What does choice programming mean to Tanzanian parents? How do Tanzanians' understandings of choice compare with and differ from U.S. missionaries' views?

To answer these questions, I propose a sociocultural model of education, a model in which institutionalized forms that appear to be similar across time and place are informed by alternative and sometimes competing registers of value and meaning. My discussion will center on two main issues. First, I shall draw out some associations between education and choice present in the words and practices of Tanzanian parents, and I shall recount some parents' arguments that "parental choice in education" has in a sense been going on since roughly the mid-1980s. Second, I shall draw out U.S. missionaries' general conceptions of choice and parental involvement, and I shall show how the idea of choice embedded in the missionaries' program begins with a very different understanding of local Tanzanian social structure. From these two points will emerge a picture of the differing histories and structures of meaning that Tanzanian parents and U.S. missionaries draw upon when discussing and designing programs that are intended to attract parental involvement. This picture will help to advance a framework that accounts for the simultaneous development of cultural variability *and* institutional homogeneity in the arena of mass education.

While my focus is on communities in northern Tanzania, my argument has general implications for other parts of the world where school choice programs are viewed as commensurate with plans to restructure education. This includes the United States and the United Kingdom, where parental involvement is emphasized as key for ensuring local management and accountability (David, Davies, Edwards, Reay, and Standing 1997; Stambach 2001a). It also includes areas represented by other chapters collected in this volume, as well as large-scale regional areas featured by sociologists in the world-culture school (e.g., Ramirez and Ventresca 1992; Meyer 1992). Common throughout the world is the interplay of local beliefs and large-scale, institutionalized forms of schooling. If parent views and school choice policies are complex and heterogeneous in Tanzania, what might be their interrelation in other cultural-geographic locations? To put this question in another way, how do transnational categories of, in this case, choice programming play out it in and "reenter" local narratives of schooling?

THEORETICAL FRAMEWORK

A few years ago, Alex Inkeles (1998) elaborated a model of convergence and divergence in the social organization of contemporary industrial societies. He argued that "there is a clear tendency for national educational systems to converge on common structures and practices," yet there "are also many instances of parallel change" and "instances in which educational systems diverge rather than converge with the passage of time" (p. 139). Inkeles' tested dimensions include ideational and legal aspects, structural aspects (e.g., grade-level structures, teacher education, formal testing), demographic aspects (enrollment ratios and repetition rates), administrative/financial aspects (ministry organizations, inspectorates, local governing boards, etc.), and interpersonal and institutional dynamics (teacher attitudes, classroom dynamics, etc.) (Inkeles 1998:14, Table 6.2). While Inkeles' model accounts extensively for the administrative and organizational aspects of schools, it does not address the many cultural contingencies that play upon people's lives and enter into schools, nor does it examine the multiple angles from which institutional forms are perceived and created—issues that would help to explain the sometimes apparent differences between global-institutional forms of schooling and local practices.

Like Inkeles, I am interested in asking, what can we observe and infer from the ideological and organizational structure of schooling? In answering this, I go further in asking, how do people's experiences, responses, and interpretations of events inform their views and expectations about education? I use anthropological arguments about structure and meaning (Dirks, Eley, and Ortner 1994) to explain why, and how, the institution of schooling remains deeply imbued with multiple and sometimes contradictory visions about the relative place of parents in education. As Masemann notes (1999:115), an anthropological approach does not constrain the researcher to consider "only the subjective experience of the participants." More dynamically, it "delineates connections between the micro-level of the local school experience and the macro-level of structural forces at the global level that are shaping the 'delivery' and the experience of education in every country" (1999:115). Delineating these connections is important for clarifying why similar programs and policies may be understood and implemented in different terms. My emphasis here is on the *differences* we see between Tanzanians' and U.S. missionaries' interpretations of choice programming in education—notwithstanding their similarities. That is, I am interested in understanding the contradictory values behind an emerging common form—particularly a growing trend toward involving parents in choices about programming in education—and in understanding how different cultural values explain the absence of complete uniformity.

It is possible that a focus on "difference" runs the risk of being misread as an interest in dwelling on "the exotic" or in overlooking the forest for the trees. It runs the risk of being understood as focusing on intricate detail to the neglect of commonalities. Yet a sociocultural approach reveals that while institutional homogeneity is indeed pervasive, and while institutional convergence has been convincingly demonstrated at the level of the form of organizations, the "forest" is made up of diverse and resilient "strands"; mass education is made up of many perspectives. My work considers, where the world-culture school generally does not (e.g., Ramirez and Ventresca 1992; Boli and Ramirez 1992), that uniformity masks underlying differences, and that understanding these differences is necessary for understanding how schools work. It is with an eye toward illuminating cultural differences that I turn—following a brief discussion of my methods and research, and a brief description of the significance of English versus Kiswahili instruction in Tanzania—to discuss parents' and missionaries' complementary and contrasting understandings of choice in education.

METHODS

This chapter is based on conversations with approximately 15 Tanzanian parents and two visiting U.S. missionaries to Tanzania (Stambach 2001b) about the state of primary schooling in Tanzania today. It is also based in a larger sense upon continuous field research I have conducted in northern

Tanzania over the past 11 years about gendered and generational transformations that Tanzanians associate with new educational opportunities (Stambach 1998, 2000) and upon two years' research on the cultural politics of school choice programs in the United States (Stambach 2001a). In my work in both the United States and Tanzania, I am generally interested in parents' interpretations of educational policies and in the relationship of educational policies to local practices and beliefs. In researching the relationship between Tanzanian secondary schooling and local household configurations in April and May 2000, I was made aware by Tanzanian parents that important changes were under way in primary education. I was intrigued by the growing presence of U.S. evangelical missionaries in providing primary education instruction to Tanzanian students, and as background to understanding the flow and transformation of educational policies across time and space, I honed in on the point of intersection between U.S. missionaries and Tanzanian parents whose children were enrolled in government primary schools. Missionaries described their work as a matter of importing a model of "choice" in education to Tanzania and modifying it for Tanzanians. Tanzanians asked whether missionaries' work was what they wanted and whether missionaries' shared their views of education.

In researching the English-language program, I pursued intensive conversations with two sets of parents and two missionaries involved in the program's implementation. One set of parents was planning to ensure that their children did *not* participate in the English-language program. The other set had sent their children to English-language primary schools and viewed English as beneficial for expanding children's post-educational opportunities. All of the parents and missionaries were familiar to one another. The sets of parents were related by marriage, and all were actively involved in a local school and church activities. Their different perspectives do not represent the range of all possible views about the subject but rather reflect a sampling of views expressed by Tanzanians about school choice and parental involvement.

Language was key to understanding some of the hesitations as well as optimism that Tanzanian parents expressed about the missionaries' intervention. During conversations with parents and missionaries, I listened for assumptions about Kiswahili as a language of instruction and about advantages and disadvantages of an English-language program in Tanzanian primary schools. I looked for evidence of parents' and missionaries' active measures to define and delimit the scope of the government's school involvement, and I compared the content and tenor of today's arguments about language and choice with what I know of the history of Tanzanian education—in particular, the fact that nineteenth-century Protestant missionaries preferred to instruct

Tanzanians in local dialects, not in "world" languages. As the following accounts drawn from my conversations illustrate, I heard contrasting views
about having a choice within the system. These differences were clearly, if not
also subtly, expressed in the ways missionaries and Tanzanians spoke of themselves as cultural versus global citizens and in the ways missionaries and Tanzanians viewed U.S. missionaries as policymakers in Tanzania.

TANZANIAN SCHOOLS AND
MISSIONARIES' INTERVENTION

The Tanzanian government provides seven years' universal primary education to all children aged seven years and older. Except for the cost of school
supplies, transportation, and in many cases, school uniforms (costs that can
exact heavy demands on poorer households), primary education is free. Of
the approximately 11,300 government primary schools located in the country's twenty regions, 24 are located in Moshi, the capital city of Kilimanjaro
Region (Ministry of Education and Culture 1999:25). Both regionally and
nationally, as well as in the two schools in which missionaries introduced the
language program, boys and girls enroll in approximately equal numbers
(1999:1). Enrollments average 350 pupils per government primary school,
and pupil-teacher ratios average 31:1 (1999:27). High student-classroom ratios (54:1) and the poor quality of school resources have prompted many
parents who can afford to send their children elsewhere to withdraw from
government schools. Over the course of the past decade, during a time of national economic restructuring and policy changes that have oriented toward
the market (Tripp 1997), the number of private, fee-based primary schools
in the Moshi area has increased fourfold, to eight.

    In part in response to the great disparities between wealthy and poor in
Kilimanjaro Region, and between those families who can afford to pay private school fees (as much as $1,000 per year) and those who cannot, Amer-
ican Protestant missionaries proposed introducing English-language
instruction into government primary schools. Their initial vision was to create a year-long, all-day program that constituted a separate class or stream;
however, staffing limitations required that they introduce a four-week intensive session instead. In view of the fact that missionaries intended to
reach a wide and socio-economically less privileged student population, their
objective was to provide English-language instruction to students in a government primary school, not to create a separate private institution.

    In 1999, an American missionary of European ancestry who had been
living in Kenya for more than ten years approached a local Tanzanian dis-

trict education office to request a permit to teach English in government primary schools. Although his request was initially denied on the grounds that it amounted to religious instruction in government schools, the missionary pursued the matter with a local education officer. This officer's own familial connections to a government official of national rank ensured that permission was duly granted and that the missionaries' work could begin in a matter of weeks. By June 1999, a group of four American missionaries and two Kenyan evangelists set up a part-time English-medium instructional program for about 200 children in Standard IV. The children in Standard IV ranged in age from 11 to 13 years old (students enroll in Standard I at age seven or older) and in terms of gender were of approximately equal numbers. The program had no prerequisite for admission other than student interest and parental permission. In 2000, the program expanded to a second government primary school, although due to special scheduling, the second school is temporarily not able to participate now. Today, in 2002, the program in the original school it is in its fourth year and continues to serve approximately 200 students during a four-week session.

To staff the program, American missionaries recruited and trained a cohort of approximately twenty American college students, themselves enrolled in Christian colleges and universities in the United States. Students trained for a few days in advance of arriving in Tanzania. They learned general facts about Tanzanian schools and about dress codes, foods, and customs in East Africa. Some were majoring in Bible Studies and training to become professional preachers and/or youth ministers. Others were students of the liberal arts and sciences at Christian colleges and universities.

American college students taught reading, art, drama, English, games, and music to the Tanzanian students enrolled in the language program. Each day, American students presented a particular lesson, such as one involving numbers and counting, and they tailored the lesson around a particular theme, such as the story of creation. As there are no formal statistics on Tanzanians' religious affiliations (although by convention most Tanzanians say that one-third of all Tanzanians are Christian, one-third are Muslim, and one-third are traditionalist), it is difficult to know how many students were familiar with the religious content of the missionaries' lessons. However, insofar as some Tanzanian teachers and parents considered that American Protestants' Bible stories might alienate some Tanzanian students, Tanzanian educators required that American missionaries use only Old Testament teachings, as a way of including at least all Christians and Muslims. While the overtly evangelical aspects of the mission were theoretically separate, Old Testament religious concepts informed the language program curriculum.

American missionary presence in government schools was, and remains, unusual. Not only were American missionaries the first religiously affiliated group (of any nationality or religion) to receive a permit to teach English in government schools; they were also one of the first groups to actively suggest changing the language of primary school instruction. Missionaries intended that the English-medium program might provide an alternative, or choice, within the Kiswahili-medium system of education in Tanzania, and that parental interest in this program might lead the government to change to an English-medium system. "Hopefully, we can persuade more people to see the benefits of English [medium] education," said one missionary, an argument also represented in the University of Dar es Salaam journal, *Papers in Education and Development* (Rubagumya 1999). English instruction was supposed to promote greater awareness of and facility within an international economic world, an intention that signified missionaries' orientation to the "world" as the arena within which school graduates would operate. As a U.S. missionary said,

> Education should move students beyond their local worlds. They [Tanzanian students] have got to use education to see more of life, not just stay here [in Kilimanjaro Region] and work on their parents' farms.

This outward movement from local setting to global world entailed a broader project involving students' parents. Whereas in the United States, policy discussions about parental choice and involvement often suggest that, given the right facts, parents will make informed and rational choices for their children, missionary discussions about Tanzanian parents' involvement suggest that parents themselves need to be instructed about how to rethink a standard model of education—in this case, the model of Kiswahili-medium primary school instruction. At various points in conversation with the missionaries, a picture emerged of the need to educate parents about the value of English instruction:

> We work through interpreters to try to tell parents that this program will make their children more international, more likely to get a job after graduation.
>   We don't evangelize in the schools. We keep the schools and our mission work separate. But we do invite parents to join us at our services.
>   It's a matter of educating the parents too. Of getting them to see the value of English for getting their kids up and out of the home.

In practical terms, parental involvement was more an ideal than a formalized aspect of the program. Parental involvement figured prominently in the missionaries' rationale for offering English instruction: Getting parents involved and giving them a choice was one of the main reasons for entering the government primary schools.

TANZANIAN PARENTS' VIEWS OF PARENTAL INVOLVEMENT AND CHOICE IN EDUCATION

Parents, however, did not uniformly accept the program and many were in general less prepared to be advocates of curricular change than missionaries had hoped. "Parents are sometimes harder to reach than students," one missionary conveyed. Many parents were not disposed to organize as committees nor, with the exception of the more experienced missionaries, did missionaries' Kiswahili language skills always facilitate clear parent-missionary conversations.

In talking with the parents, I quickly noticed a difference of opinion about the relative autonomy of parents in matters of school administration, and I noticed that despite variations, Tanzanians generally considered the missionaries' program in the context of their understanding of postcolonial history, while the missionaries emphasized that English instruction was a sound economic strategy—a "rational choice" in an era of world markets. The parents generally supported Kiswahili as a language of primary school instruction, and they expressed a general affinity for the African qualities they associated with Kiswahili. The missionaries, however, emphasized the importance of English as a way of opening up new economic and social opportunities for Tanzanian children. These different views resonated with different assumptions about the meaning and history of mass education, particularly the place of national language instruction in preparing students for life after graduation. While all of the missionaries and parents agreed that English was a language of international communication, the Tanzanian parents were more apt to value Kiswahili as a language that promoted cultural cohesion. The missionaries, in contrast, portrayed Kiswahili as a culturally unique but economically stymieing language for the region. To the parents, Kiswahili was the nation's mother tongue, while to the missionaries it was a language of limited practical use.

Among the Tanzanian parents who intended to withhold their children from the program, one set described the English-language program as "another form of Western imperialism." One father—himself a college-educated businessman who spoke English as his third language—asked,

> Why should we choose English in primary schools? Kiswahili is our national language. English comes at the secondary level. . . . How are our

children going to communicate if they don't learn how to speak proper Kiswahili at school? . . . English will come. Students will learn English in due time. But for primary school, they should learn their African tongue.

His argument supported and reinforced thirty years of official Tanzanian language policy. Kiswahili is the official language of instruction in government primary schools. It remains the primary language of communication throughout secondary and tertiary levels of education. With more than 135 local dialects spoken within the borders of Tanzania, Kiswahili serves a practical and symbolic function of uniting the country under a single African language. While some Tanzanians sometimes view Kiswahili impatiently as an instrument of state authority, or as the cultural property of coastal Swahili people for whom Kiswahili is their first language, Kiswahili is also widely regarded as a unifying symbol that reflects cultural peace and social stability. Some parents' reluctance to embrace the English-language program suggests a more subtle reluctance to divest from supporting Kiswahili as a unifying cultural symbol.

The missionaries' answer to the parents' concerns for preserving Kiswahili was expressly that "English provides Tanzanians with a choice in education." To paraphrase one of the founding missionaries:

Our language course gives parents the choice of educating their children in a world language and of having a future that moves them beyond Tanzania. . . . Kiswahili is important and wonderful, but where are they going to use Kiswahili outside of parts of East Africa? . . . If these kids want to have any hope of getting into business and getting out of this poverty, they have to know English, and they have to be able to speak to the rest of the world in English. . . .

The Tanzanian parent paraphrased above did not completely disagree with the missionary, nor did his wife, who responded to the missionary's line of reasoning by saying:

Of course English is the language of international commerce and business. And of course I want my children to learn English. . . . But Kiswahili has been important to our country. Schools are the places where our children learn standard Kiswahili. Like me, they can learn English later if they need to.

This set of parents—the set opposed to the language program—generally felt that educators, not parents, should take charge of education, and that their children's educators should be Tanzanian. They questioned whether parents' decisions about language instruction were as well informed as professional educators'. Instead of starting from the standpoint that parents necessarily know what is best for their children, they considered that Tanzanian teachers and official curriculum developers would have a better sense of what Tanzanian school children ought to study and know.

In contrast to these views were those of the other set of parents, who subscribed to the missionaries' ideals of English instruction yet shied away from what they regarded as the missionaries' "outsider" approaches to education policy. These parents questioned the novelty of the ideas of parent involvement and choice in education and argued that parents had long been involved in the organization and administration of primary education. Over the course of roughly the past fifteen years, since the early days of the state's shift toward a policy of seeking private investment to fund public education (particularly at the secondary level), parents have frequently played important, if largely unofficially recognized, roles in both primary- and secondary-level school fundraising, staffing, and to a lesser extent, curricular development. Core groups of parents have long been active in developing alternative programs, including special programming for students in the areas of carpentry, handiwork, and career development, and in some cases for English-language instruction. To paraphrase a Tanzanian parent who was sending her first- and second-born children to a private English-medium primary school and who, in addition to helping to organize a privately funded village secondary school, had been a member of the parents' committee at the local government primary school:

> There is a difference between people coming from outside and people pushing for changes from within. One difference is that outsiders are sometimes called big bosses [in Kiswahili, *mabwana;* also sometimes called "colonials," *wakoloni*]. Insiders are more likely to be called friends and to get the support of parents in general.

I asked this parent why nonetheless so many Tanzanians seemed to support foreign investment and international development, particularly in the northern region, where development projects were plentiful and communities often competed to attract assistance. Her response, that "we're more interested in the bulldozers and Land Rovers left behind than in the projects themselves," had implications for the English-language program. For it seemed that the English-language developers had come to Tanzania with what she viewed as valuable technology and organizational skills. An unintended but desirable benefit of this program, this woman said, would be the likely introduction of several personal computers and knowledge about how to start a new English program, including what materials to develop and how to go about fitting a new program into schools. Critics of the language program shared this view: The college-educated businessman above, who questioned the language program on the basis of its seeming imperialism, argued that the resources brought, and sometimes left, by missionaries were valuable contributions to Tanzanians and government schools. But unlike him, parents who supported the English-language program did so because they thought English was a desirable and necessary skill; it was not only the materials and organizational skills but also the language training they found attractive.

In one respect, the argument advanced by this second set of parents sounded much like the missionaries' pragmatic argument. These parents reasoned that the language program presented parents with new opportunities and hence was not a program they could reasonably reject. However, this second set of Tanzanian parents questioned the novelty of trying to bring parents into the system. The mother who anticipated material gains went on to say,

> Parents have been involved in educational programming for many years, but with the exception of *wazazi* schools [established in the 1970s under the auspices of the then sole political party, Chama cha Mapinduzi], parents have never officially been involved in government primary education. Yet we've taken charge of primary schools at times when there's been no internal organization, and we've used our own money and money we've raised from other sources [including international church organizations]. . . . I guess what's different now is that parents are being brought into decision making more formally. Or at least that's what it looks like they're trying to do.

This parent suggested that government-sponsored primary education was and continues to be prone to failure. Mismanagement and/or lack of funding were recurring themes, she said, in government education. More successful have been parents' own community-based interventions and efforts to tailor primary school programs to the needs of communities' children—an argument remarkably similar to choice advocates' arguments in the United States for community-based educational interventions, but different in terms of the history of the state and, by extension, choice advocates' orientations to government. Whereas choice advocates in the United States (among whom the missionary counted himself) regard the government as being generally "of, by, and for the people," the history of colonialism in Tanzania continues to flavor the state as a governing body "from outside," particularly when it relies heavily on foreign aid and investment and on internal consultants such as this team of American missionaries.

It is important to remember that this parent and her husband were not critical of the concept of English-medium instruction. Like the missionaries, they believed that strong command of English was essential for students' mobility. Instead, they questioned the efficacy of foreign missionaries in the schools and wondered whether government bureaucracy would not weigh down a formally organized system involving parents. Their concern about over-bureaucratization appears superficially to be remarkably similar to choice advocates' arguments in the United States, where advocates push for "maximum flexibility" in states' policies on choice programs so that parents' involvement remains minimally encumbered (Center for Education Reform 2000). However, to regard this commonality as an indication of a world institutional trend overlooks the different contingencies that play upon people's lives (such as Tanzanians' need for technological resources and the historical forces that return U.S. missionaries to Tanzania) and that directly inform people's actions (such as Tanzanians' decisions to support the missionaries' choice program but not for all of the reasons the missionaries give). It is these contingencies that matter to any final explanation of why forms appear similar but operate differently.

U.S. MISSIONARIES' VIEWS OF PARENTAL INVOLVEMENT AND CHOICE IN EDUCATION

To return to the side of the U.S. missionaries, and to a sociocultural analysis of the context within which the missionaries propose their English-language program to Tanzanians, it becomes necessary to step outside the immediate logic of the missionaries' arguments and examine the assumptions upon which they rest. The missionaries' introduction of choice programming into Tanzania needs to be understood within the context of two assumptions that thread through American society: one, the idea that foreign aid, particularly in the form of education (and specifically English-language instruction) can alleviate poverty and provide access to world markets; and two, an assumption that parental involvement in choice options in education will result in expanded post-educational opportunity and changes within the government system. In this section, I will demonstrate how these assumptions manifest themselves in the missionaries' work in eastern Africa, and I will discuss how they resonate with Tanzanians' experiences.

It is perhaps easiest to illustrate a belief in the idea that foreign aid in the form of language instruction can alleviate poverty by repeating the message expressed in the missionary's comments above about the value of the English-medium language program: that the English-language program has the potential to give "parents the choice of educating their children in a world language" and to enable Tanzanian students to get "into business" and "out of this poverty." The missionary's message evinces a belief in an updated version of human capital theory. As Patrick Fitzsimons (1994, 1999) discusses, human capital theory is one of the most influential economic theories of public education. It operates according to a belief that education maximizes an individual's economic potential and that the economic health of a nation is the sum of its citizens' productivity. In an updated version, human capital theory is reformulated to produce a new role for education (OECD 1997); education is now seen as having the potential for moving individuals beyond the realm of the nation-state. This updated version stresses that education and training are "key to participation in the new global economy" (Fitzsimons 1999) and emphasizes that investment in financial markets and electronic skills take priority over skills that contribute to the manufacture of commodities.

The missionaries' belief that English-language instruction can alleviate poverty also rests on an assumption that people always act on the basis of their own economic self-interest. This assumption informs human capital theory and is modeled around a particular notion that individuals act singly and with relatively few constraints put on them by others. The missionaries' expectation is that parents will choose English instruction for their children for the reason that their children's command of the language will further parents' own interest, as well as that of their children. However, human capital theory has been criticized for dismissing the pull of social groups on individuals' decisions and for assuming that all decisions are made on the basis of maximizing economic self-interest (Block 1990). If the set of Tanzanian parents who were concerned that English would erode the symbolic value of Kiswahili is any indication, some parents are not always operating according to principles of maximizing self-interest but on principles that stress the maintenance of group ties.

The missionaries' belief that parental involvement in choice options in education will result in expanded post-educational opportunities reflects an assumption that individuals operate freely within competitive markets and that competition maximizes individuals' opportunities. In contemporary human capital theory, "human behaviour is based on the economic self-interest of individuals operating within freely competitive markets. Other forms of behaviour are excluded or treated as merely distortions of the model" (Fitzsimons 1999; see also Fitzsimons 1994). State programs, such as publicly financed systems of education, are seen from this viewpoint as artificially controlled and noncompetitive. Human capital theorists—largely those who contribute to today's educational policymaking—regard centralized state education systems as undesirable on the grounds that they are bureaucratically top-heavy and inefficient. Choice in education becomes desirable from within the human capital framework for creating more choices and generating competition that, in turn, improves the main system of public education.

In the United States, such thinking lies behind a growing trend toward choice programs in education. The "economic fallacy" of the argument—the misconception that economics resides outside of culture (Marginson 1993:25 cited in Fitzsimons 1999)—is made all the more apparent in non-U.S. settings where markets, education, and the concept of choice have all been layered onto historically and culturally non-Western understandings of transaction and personhood and imbued with alternative registers of meaning. The concept of choice and the logical machinations of markets in East Africa are complex sociocultural phenomena. They play into a dynamic of interpersonal relations that are sometimes expressed in terms that are very foreign to Americans (see, for example, Ciekawy 1998, on sorcery and witchcraft), and they emerge within a historical context that sometimes associates education with Western power. The missionaries' original thinking—that the English-language program would attract a large group of Tanzanian parents looking to have a "say" in their children's education and looking to implement an English-medium government primary school—took a turn in direction toward a more missionary-led enterprise when only a handful of parents spoke in support of an English medium primary school and only a very few viewed the program as a form of choice. Whereas many parents sent their children to the four-week program with the hopes that their children's English-language skills would improve, none worked actively to support the program and none linked it to a wider English-language movement for school choice. The mother in the second set of parents above noted that

> parents know a good thing even though they know it's flawed. They're not going to *not* enroll their children in the program when they know their children could get something out of it. But neither are they going to endorse it wholesale.

Parental involvement remained an ideal, but in reality, the program was missionary led.

That some parents had questions about the language program may be an indication that the cultural calculus at play among some Tanzanians was not one of freely competitive markets but of markets differentially controlled by some groups and not by others. Such is certainly a consideration that the second set of parents had in mind when they said that Tanzanian parents, not visiting missionaries, should set the parameters of language instruction. For between the degree to which the missionaries took charge of the project and the context within which the Tanzanian parents interpreted and even rejected the concept of parental involvement in education, the English-language program was hardly a replica of choice programming in the United States. It interfaced with a historically and culturally very different system in Tanzania and with a different history of parental involvement in education. Yet on the surface, missionaries and local officials continued to call the program a "choice" and continued to encourage the teaching of the English language—two aspects that might suggest a convergence in form and a move toward an overarching world system of education. In the final section, I will consider the deeper question of the directionality of institutional forms in education. In particular, I will ask whether an emphasis on interpreting commonality of institutional forms is not itself a reflection of a core cultural dynamic—or, for that matter, if an emphasis on "difference" is not also a part of a larger cultural creation.

WORLD-CULTURAL AND ANTHROPOLOGICAL VIEWS OF PARENTAL INVOLVEMENT AND CHOICE IN EDUCATION

Since at least the mid-1950s, with the rise of studies in comparative education, an ongoing dichotomy of views has prevailed (Kazamias and Schwartz 1977): that of whether local practices influence institutional forms or whether institutional reforms reflect a universal goal and, in turn, shape local practices. Cultural anthropologists tend to weigh in with the former point of view; world-culture sociologists tend to support the latter. Yet, as Andrew Strathern argued several years ago, the two views logically entail one another:

> Locally interpreted narratives give force to . . . universal categories, yet these same categories reenter the narratives as [a] means of interpreting them. The movement of thought is thus not simply linear, but also recursive or circular at certain points. (Strathern 1995:178)

Any consideration of transnational educational policies needs to account for this recursive dynamic (cf. Kinoshita 2002). The task is not one of identifying what is universal or converging, nor to label and minutely specify what is unique about each situation, but to address how locally interpreted narratives give force to universal categories *and* how universal categories give force to local narratives. In essence, universal categories are historical and "transnational"; they do not exist uniformly everywhere nor do they operate under the same conditions. Categories of choice and parent involvement, for instance, are carried from one locale to another and are imbued with meanings that are sometimes incommensurate.

In Tanzania, the case of missionaries promoting parental choice in government schools illustrates how transnational categories of choice and human capital theory catalyze local narratives about economy and parental involvement. Tanzanian parents' views of history, of aid, and of Western imperialism are tied to colonialism, to local registers of economy (Ciekawy 1998), and to particular positions and perspectives on a world system. Missionaries' views of "choice," though introduced as a universal frame, are tied, in contrast, to American-historical ideals of freedom and autonomy and to beliefs about the rights of individuals to have a say in education—ideals and beliefs that play out differently in local settings in Tanzania and the United States. Missionary choice proposals lend support to Tanzanian arguments that Kiswahili *is* a choice. Missionary proposals to maximize student opportunity by involving parents in education raises awareness among Tanzanians that Tanzanian parents have long been involved in education and that some aspects of life, such as social unity derived through the symbolic value of Kiswahili, are worth more than individuals' outward, global mobility.

Examined from another direction, that of how Tanzanian narratives inform and partly alter the missionaries' ideas and programs, the local narratives of some Tanzanian parents reinscribe missionary categories of choice within a more contextualizing, less universalizing framework by showing that choices are not the same everywhere. If choice is truly an open construct, then choices will be variable. Instead of choosing to school their children in English-medium primary schools, some parents now choose to continue with Kiswahili. Others choose to enroll in the program but contend that parent involvement in educational programming is not new. In these ways, the choices the missionaries introduce reenter Tanzanian narratives as a means of making and interpreting educational policy differently.

The uneven reinscription of local beliefs and practices is key to understanding why cultural forms appear to be isomorphic. The missionaries' particular registers of economy and ideals of freedom (which rest, as it were, on internationally dominant theories of human capital) are more likely than some Tanzanians' to be taken up in institutionalized forms of schooling. Uneven access to and control of resources leads to some beliefs gaining stronger footing in the category termed "universal." This is why similar forms appear; but similar forms do not mean cultural uniformity.

For the present chapter, it is important to note that even from a world-cultural perspective in which institutional forms are seen as universal and converging, choice programming in Tanzanian education is a tenuous phenomenon. Within the limited scope of the English-language program I have described here, the program has all the characteristics that any typical choice program might have in the United States: a stated commitment to parental involvement and institutional flexibility, and an institutional structure that merges private organizations (the mission and its summer youth ministry program) with public institutions of education (the Tanzanian system of government primary schools). Yet the program attracts parents who note that they already have a history of involvement in their children's education and who are aware of larger cultural contradictions between the government system of public education and their cultural beliefs and practices. It is also implemented less as a choice that is parent driven than as an intervention that is mission led.

In view of the complexities of introducing choice programming into the system of education in Tanzania, it might be reasonable to answer one of this volume's overarching questions—What is the significance of a global culture if it exists at all, even in the most superficial form?—by saying that to the U.S. missionaries involved in this language project, the significance is very great: A global culture is what promises to move Tanzanians from poverty, and the freedom and opportunity to choose English-language instruction is a precursor to that. To Tanzanian parents, the answer might be that a global culture threatens to erode something intangible they identify as "culture" and that they see is present in Kiswahili. The parents discussed here are not any more "particularistic" in their cultural views than the missionaries are "universalistic." That is, even though the missionaries' views are more likely to be taken up in the institutionalized forms of schooling, both missionaries and Tanzanians have localized visions of universal forms of schooling. In fact, as described, the transnational category of choice in education is itself tied to local understandings of economy and interpersonal relations. Both missionaries and Tanzanians express an understanding of local differences and universal categories. If anything might be noted as different between them, however, it would be that the missionaries work for an unquestioned view of the global, while the Tanzanian parents clearly express an alternative to the missionaries' belief in English and in choice programming as a universal.

ACKNOWLEDGMENT

This chapter is based on fieldwork funded by the Spencer Foundation and the University of Wisconsin–Madison. I am grateful to the Tanzania Commission for Science and Technology (COSTECH) for granting me permission to conduct research in Tanzania and to Kathryn Anderson-Levitt and Stacey J. Lee for their helpful suggestions in preparing this chapter for publication.

REFERENCES

Ball, S., R. Bowe, and S. Gewirtz. 1995. Circuits of schooling: A sociological exploration of parental choice of school in social class contexts. *The Sociological Review* 43: 52–78.

Berman, E. H. 1999. The political economy of educational reform in Australia, England and Wales, and the United States. In *Comparative education: The dialectic of the global and the local,* edited by R. F. Arnove and C. A. Torres. Lanham, MD: Rowman and Littlefield.

Block, F. L. 1990. *Postindustrial possibilities: A critique of economic discourse.* Berkeley and Los Angeles: University of California Press.

Boli, J., and F. O. Ramirez. 1992. Compulsory schooling in the Western cultural context. In *Emergent issues in education: Comparative perspectives,* edited by R. F. Arnove, P. G. Altbach, and G. P. Kelly. Albany: State University of New York.

Bray, M. 1999. Control of education: Issues and tensions in centralization and decentralization. In *Comparative education: The dialectic of the global and the local,* edited by R. F. Arnove and C. A. Torres. Lanham, MD: Rowman and Littlefield.

Center for Education Reform. 2000. About School Choice. http://edreform.com/school_choice/

Ciekawy, D. 1998. Witchcraft in statecraft: Five technologies of power in colonial and postcolonial coastal Kenya. *African Studies Review* 41 (3): 119–41.

David, M. E. 1993. *Parents, gender, and education reform.* Cambridge, MA: Polity Press.

David, M. E., R. Edwards, D. Reay, and K. Standing. 1997. Choice within constraints: Mothers and schooling. *Gender and Education* 9 (4): 397–410.

Dirks, N. B., G. Eley, and S. B. Ortner. 1994. Introduction to *Culture/power/history: A reader in contemporary social theory,* edited by H. Dirks, G. Eley, and S. Ortner.

Fitzsimons, P. 1994. Human capital theory and the government's industry training strategy. *Journal of education policy* 9 (3): 245–266.

Fitzsimons, P. 1999. Human capital theory and education. *Encyclopedia of Philosophy of Education Online.* http://www.educacao.pro.br/humancapital.htm

Good, T. L., and J. S. Braden. 2000. *The great school debate: Choice, vouchers, and charters.* Mahwah, NJ: Lawrence Erlbaum Associates.

Inkeles, A. 1998. *One world emerging? Convergence and divergence in industrial societies.* Boulder, CO: Westview Press.

Kazamias, A. M. and K. Schwartz. 1977. Intellectual and ideological perspectives in comparative education: An interpretation. *Comparative Education Review* 21 (2 and 3): 153–176.

Kinoshita, U. 2002. Anthropology of policy: Strengths and weaknesses. Manuscript, Department of Educational Policy Studies, University of Wisconsin–Madison.

Marginson, S. 1993. *Education and public policy in Australia.* Cambridge: Cambridge University Press.

Masemann, V. L. 1999. Culture and education. In *Comparative education: The dialectic of the global and the local,* edited by R. F. Arnove and C. A. Torres. Lanham, MD: Rowman and Littlefield.

Meyer, J. W. 1992. World expansion of mass education, 1870–1980. *Sociology of Education* 37 (4): 454–75.

Ministry of Education and Culture. 1999. *Basic Education Statistics for Tanzania.* Dar es Salaam, Tanzania.

OECD. Organisation for Economic Co-Operation and Development. 1997. *Internationalisation of higher education.* Paris: Centre for Educational Research and Innovation.

Ramirez, F. O., and M. J. Ventresca. 1992. Building the institution of mass schooling: Isomorphism in the modern world. In *The political construction of education: The state, school expansion, and economic change,* edited by B. Fuller and R. Rubinson. New York: Praeger.

Rubagumya, C. M. 1999. Choosing the language of instruction in post-colonial Africa: Lessons from Tanzania. *Papers in Education and Development* 20: 125–145.

Stambach, A. 1998. Education, mobility, and money: Reflections on the cultural meaning of educational investment. In *Advances in educational policy, volume 4,* edited by K. Wong. Greenwich, CT: JAI Press.

Stambach, A. 2000. *Lessons from Mount Kilimanjaro: Schooling, community, and gender in East Africa.* New York: Routledge.

Stambach, A. 2001a. Consumerism and gender in an era of school choice: A look at US charter schools. *Gender and Education* 13 (2): 199–216.

Stambach, A. 2001b. Education and Christian fundamentalism. Paper presented at the University of Chicago Conference, Rethinking the interrelationship of anthropology and education. May 2001.

Strathern, A. 1995. Universals and particulars: Some current contests in anthropology. *Ethos* 23 (2): 173–186.

Tripp, A. M. 1997. *Changing the rules: The politics of liberalization and the urban informal economy in Tanzania.* Berkeley and Los Angeles: University of California Press.

CHAPTER SEVEN

THE POLITICS OF IDENTITY AND THE MARKETIZATION OF U.S. SCHOOLS
How Local Meanings Mediate Global Struggles

Lisa Rosen

This chapter seeks to develop a locally grounded understanding of the advancement of neoliberal approaches to government within the arena of public education. The discussion analyzes a case of local debate over the "marketization" of education in one community as a focus for illuminating the relationship of translocal struggles over political and educational restructuring to particular, localized conflicts. I examine the processes of cultural production at play in this debate, in which a group of parents used a neoliberal vocabulary to argue for parental choice of curriculum materials and instructional methods in their children's classrooms. This group represented themselves as "customers" of schools and argued for their entitlements to "consumer choice." This claim was vigorously contested by an opposing group of parents and teachers who challenged the legitimacy of the "customer" identity and of descriptions of schools in the idioms of the market. The discussion considers the clash between competing modes of representation—particularly the politics of identity surrounding the characterization of parents as the "customers" of schools—as a focus for examining the relationship between processes of local debate and more enduring changes in the political economy of education.

Specifically, I argue that translocal struggles do not simply determine local conflicts; rather, local people appropriate them for their own purposes, and local conflicts partially shape translocal struggles. In this case, local parents appropriated the neoliberal language of "choice" out of frustration at their inability to influence local curriculum decisions; the success of this discourse strategy had broader consequences, both for the school district and perhaps also for translocal struggles over educational authority and governance. The discussion demonstrates that large-scale structural transformations such as the global spread of neoliberalism are partly constituted by local processes of struggling that can produce enduring effects. At the same time, local struggles also partake of and are partly constituted by translocal structures and by struggles over these more enduring structures. In starting from this premise—that specific local conflicts and more long-term struggles are mutually constitutive—I aim to contribute to the theoretical project recently articulated by Holland and Lave in *History in Person: Enduring Struggles, Contentious Practice, Intimate Identities* (2001). In particular, my analysis builds upon their argument that local struggles are key sites for the cultural production of identity, and that these constitutive processes—and the politics surrounding them—mediate between local conflicts and more enduring struggles.

In adopting this theoretical stance, I also present an alternative to how the global spread of neoliberal policies has generally been conceptualized within the literature on recent educational reform.[1] With some exceptions (for example, Lingard 2000; Luke and Luke 2000), this literature has tended to interpret the increased salience of such policies as "part of a broader economic, political and cultural process of globalization" that increasingly shapes local practice around the world, and "in which . . . state bureaucracies fragment and the notion of mass systems of public welfare, including education, disappears" (Whitty, Power, and Halpin 1998:31). In contrast to this more deterministic view, this paper analyzes such changes from the "bottom up."

BACKGROUND TO THE DEBATE

The setting for this case is the town of Shady Hills,[2] a community in California known for its abundance of material and intellectual capital deriving from its proximity to both a prominent research university and a thriving high-tech sector. The student body of the Shady Hills Unified School District ranges from children of the intellectual and professional elite (from whose ranks were drawn several participants in the choice debate) to less privileged children from the poorer end of town. However, the debate in question primarily engaged a subset of the district's most involved, upper-middle-class parents.[3] The schools in the district enjoy a reputation for providing an exceptionally high quality of public education and for being on the cutting edge of educational improvements. They provide a variety of educational options (for example, a foreign language immersion program and several "alternative" elementary schools) to students in the district, who numbered just over 7000 at the time of the debate.

Since the early 1990s, the district has become engaged in the statewide movement to "reform" elementary, middle, and high school mathematics education by instituting dramatic changes in curriculum and instruction. The statewide reform agenda encompasses changes in the entire curriculum, from the teaching of early number concepts to calculus. Mathematics is traditionally taught by presenting students with a mathematical rule, illustrating its use with a sample problem, and having students imitate the technique by practicing to individual mastery on a series of similar problems. Reformers argue that all of these aspects of traditional curricula and methods must change. They say that the traditional methods of teaching mathematics have deprived students of conceptual understanding, alienated children, and failed to teach them to meaningfully apply mathematical concepts to problems and tasks outside of school. They instead promote curricula that emphasize problem-solving, mathematical reasoning, and application of mathematical concepts to lifelike situations. Reformers argue that students learn not by being given rules to follow but by "constructing their own knowledge." The preferred method is therefore "discovery learning," which may involve students working together in groups on complex tasks that engage their prior knowledge, integrate numerous concepts, and may take several days or weeks to solve. Reform programs also include collaborative writing and graphic display tasks in which students are asked to explain their reasoning and represent their solutions to the teacher and the class, not only numerically, but also verbally and/or visually (see California Department of Education 1992).

As new mathematics textbooks and other instructional materials began to appear in schools and classrooms around the state, parents started becoming concerned. Local protests began erupting around the state as concerned parents joined together to voice their alarm. However, these efforts were met by opposition from defenders of the new curricula, who formed organizations of their own. The choice debate to which this paper is addressed was a skirmish in this broader battle.

My analysis is informed by 18 months of research on the ongoing debate over mathematics "reform" in California. My discussion draws on data derived from: participant observation at public meetings throughout the state, interviews and discussions with key players statewide, and analysis of a wide variety of texts relevant to the debate, including news reports, minutes from public meetings, public testimony, editorials, columns, letters to the editor from various newspapers, and other correspondence. The discourse strategies with which the chapter is concerned are illustrated primarily by examples from documents and correspondence (both published and unpublished) from the period in which the choice debate occurred, from January to June 1997. My analysis also draws on conversations and interviews (conducted in early 1998) with participants in this local debate.

THE CHOICE DEBATE

In November of 1994, a group of Shady Hills parents created an organization—HOLD (Honest Open Logical Debate on math reform)—to protest the changes in mathematics education. They argued that the new approach was insufficiently rigorous to provide their children with a solid mathematics education and pointed to a recent drop in local test scores as evidence for this charge. Members of HOLD immersed themselves in the national literature on educational reform, publicized their criticisms of the reform movement and the local math program, sought support from parents, scientists, and politicians in other cities and states, brought camera crews into teachers' classrooms, protested outside PTA events, joined school-, district-, and state-level educational committees, and communicated their concerns to the school board through letters and public testimony. However, HOLD's attempts to return the district to more traditional methods were opposed with equal vigor by a local organization formed in May of 1995 to defend teachers and the mathematics program from attacks by HOLD: Parents and Teachers for a Balanced Math Program (also known as Pro-Balance, or the Pro-Balance coalition). They charged that HOLD's efforts were not only an attempt to return the district to an outmoded educational philosophy, but also an attack on the professional authority of the district's competent, dedicated teachers.

After more than two years of aggressive campaigning, HOLD had still not succeeded in their goal of ridding the district of "reform" math. Their media campaigns and other organizing efforts had been strenuously opposed by Pro-Balance, who had succeeded in making as much noise in favor of the reforms as HOLD had made against them. In January of 1997, a few members of HOLD got together with some like-minded district parents and formed a new strategy. They revived an idea they had first suggested in March of 1995: the establishment within each middle school of separate "houses" (which they now referred to as "teams") that would offer different educational approaches. Presenting this earlier idea in a new vocabulary—that of the market—they formed a new group to pursue this agenda exclusively. Calling themselves the Parent Committee for Sixth Grade Choice (hereafter referred to as the Parent Committee), they requested that the Board of Education offer a "Direct Instruction Team of Choice" at both of the district's middle schools, emphasizing four principles:

- the acquisition of academic skills and a "knowledge-based curriculum,"
- a consistent educational philosophy of clear expectations and a quiet, orderly environment,
- proven academic achievement as evidenced by quantitative measurement of student progress through testing and grades, and
- strong connections with parents though active parent influence—"the parent is the customer."

Specifically, they asked the school board to offer sixth-grade students a "direct instruction" teaching alternative for the "core subjects" of math, language arts, social studies, and science. Students in the direct instruction team would be grouped together to receive separate instruction for the latter subjects, but would mingle with the rest of the sixth grade for their other three classes (such as music, physical education, Latin, and home economics).

The Parent Committee presented their proposal as a plan to "restore harmony to the district." The years of conflict leading up to their proposal had left scars on all sides. For example, members of both HOLD/The Parent Committee and Pro-Balance shared with me in later interviews their feelings of being frustrated by the conflict and disrespected by their opponents. The community had seen experienced teachers transfer or quit in disgust and frustration over the controversy, while others became fearful and defensive. Members of HOLD and Pro-Balance, meanwhile, became increasingly resentful and mistrustful of each other's activities, a dynamic that spilled over into discussions of the controversial choice plan.

Pro-Balance immediately mobilized against the choice proposal, seeing it as an attack not only on the district's beleaguered teachers and their much-maligned instructional program, but also on the collective good of the community. Their concerns were shared by parents and teachers unaffiliated with either HOLD or Pro-Balance but alarmed by the implications of the plan—the remaking of the public schools in the image of the market. The ensuing debate lasted for six months and received coverage in both the local and regional print media, as well as on the World Wide Web. In letters to the media and public testimony to the school board, Pro-Balance challenged each of the Parent Committee's claims about education in the district, as well as the general notion that parents are the "customers" of the schools. Meanwhile, the Parent Committee gathered signatures from the parents of incoming middle school students in support of their proposal, which they presented to the school board as evidence of widespread interest among parents in having their children placed in a direct instruction classroom. The board agreed to explore the possibility of revising its policy on alternative programs, in order to establish guidelines for the possible creation of "teams of choice." Perhaps encouraged by this willingness to consider accommodating the Parent Committee's proposal, a third group of parents two weeks later presented the board with a choice proposal of its own. They asked the board to create a third option, a "Collaborative Learning Team" that would emphasize cooperation, student-directed study, in-depth learning, community service, and creativity.

Two months and numerous revisions later, the Board of Education adopted a new policy on alternative programs. The new policy specified the conditions under which a choice program might be implemented, which were intended to address the concerns of the opponents of choice.[4] The board then surveyed parents in the district to determine their instructional preferences, which resulted in the creation of one collaborative learning class at one of the two middle schools and one direct instruction class at each of the two schools. According to a Pro-Balance member, at the meeting at which the decision was announced, a leader of both the Parent Committee and HOLD presented the board with a sheet cake: chocolate on one side, vanilla on the other, and both flavors in the middle. With great ceremony, he presented each board member with a slice, congratulating them for their efforts in creating the "middle school choice program."

GLOBAL STRUCTURING STRUGGLES

In some respects, the choice debate was a local expression of long-standing struggles, not only over competing models of government and the different conceptions of schooling they imply, but also over questions of educational authority and expertise. However, the manner in which these more enduring struggles structured local debate did not simply reproduce long-standing struggles on local ground. Rather, local actors used a variety of discourse strategies, some of which were directly appropriated from these broader struggles, but others of which had origins elsewhere (for example, in the discourse of developmental psychology). This ensemble of resources together shaped the way in which the debate played out locally and its potential impact on translocal struggles. In the following section, I briefly summarize the long-term, interrelated struggles that both provoked and enfolded the choice debate in Shady Hills. Subsequent sections elaborate how these global struggles were appropriated locally and examine the "two-way generative traffic" (Holland and Lave 2001:22) between these enduring struggles and processes of local conflict.

Struggle over Models of Government

The contested idioms mobilized within the choice debate were appropriated from a broader struggle over models of government, "between an old social democratic model—based on a paternalistic, bureaucratic, welfarist approach to government—and a neoliberal model in which the power of government is mediated and disguised by *laissez faire* economics and flanked by an ethos of individualism" (Shore and Wright 1997:28). Contestation in Shady Hills over the legitimacy of how the Parent Committee characterized schools and parents was structured by the central themes of this broader struggle, particularly the neoliberal themes of choice and consumer rights, on the one hand, and the social democratic theme of expert administration of the public interest, on the other.

As several chapters in this volume elaborate, neoliberal philosophy has manifested in education primarily through a dual agenda of *deregulation/decentralization* and *privatization*. This involves a range of initiatives to increase autonomy from state control and provide more choices in relation to educational matters for districts, schools, parents, and students. Advocates of privatization and deregulation argue that such plans will increase competition among schools and maximize choice for consumers, forcing schools to innovate and to optimize their performance, while equalizing access to quality education by empowering parents to maximize their children's individual opportunities. These efforts, and the logic underlying them, have been staunchly challenged by a range of critics who fear that the deregulation of public education will exaggerate existing social inequalities. A key difference between the two models of government is that the neoliberal model implies that education is a "private good" (an exclusive possession, akin to other "commodities," which provides primarily individual benefit), while the social democratic model conceptualizes education as a "common good" that confers shared, collective benefits on society as a whole. With respect to the latter notion, the social democratic model assumes that schooling should serve societal interests by instilling in students the particular skills, values, and/or dispositions that are collectively deemed to be of political or social importance. According to this view, and in marked contrast to neoliberal views, government has a legitimate role in regulating the provision and content of public education in accordance with broader social goals—for example, by imparting those skills and values deemed necessary for the perpetuation of democracy (cf. Labaree 1997:5).

Struggle over Educational Authority

Contestants in the choice debate also appropriated language and ideas from an enduring struggle over the status and authority of the teaching profession and the field of education more generally. The establishment of education as a distinct field of study early in the previous century, the concurrent production of specialized vocabulary and knowledge as the basis of professional expertise, and the development of separate university-based schools of education, professional organizations, and journals created a degree of both isolation and unequal power between education experts and members of the lay public. The principles and commitments of educational progressivism, in its various incarnations (e.g., administrative, pedagogical, and/or social, in varying degrees within different institutions and at different times), came to dominate schools of education in the United States.[5] This powerful, albeit contested, intellectual tradition, and indeed the very existence of education as a distinct field of scholarly inquiry and professional training, presupposes that teaching and learning are complex processes requiring specialized knowledge and training on the part of educational practitioners. This perspective implies a more complex view of curriculum and instruction, and thus a greater expectation of *professional expertise* on the part of classroom teachers, than the "transmission" model of teaching and learning that has dominated the popular imagination for most of the last two centuries and persists into the present.[6]

In opposition to this model of teachers' work, movements of professional educators have long advocated the "professionalization" of teaching, that is, giving teachers greater authority over a range of educational decisions, elevating the status of the profession, and consolidating and standardizing the body of technical knowledge that comprises teachers' expertise. However, professionalization often comes at the cost of undermining connections and relationships between schools and communities/parents, particularly when teachers and education experts use their professional authority "as a weapon to further distance parents and communities from attaining a meaningful voice in school affairs" (Zeichner 1991:367). Indeed, since the establishment of education as a distinct professional field, public campaigns for changes in curriculum have tended to pit parents and academics favoring a "traditional" or transmission approach to curriculum and instruction against experts trained and/or housed in schools of education and advocating progressive reform. The previous century witnessed several waves of revolt in response to efforts by professional educators to reform public school curricula in keeping with progressive ideology. Common to both earlier and current antiprogressive campaigns were not only similar concerns for discipline, order, and academic excellence, but also a shared disdain for professional education experts (Cremin 1964).

LOCAL STRUCTURING PRACTICES

The following section elaborates the manner in which contestants in the choice debate marshaled cultural resources to press their claims and challenge the claims of their opponents. In particular, I analyze the rhetorical structure of the debate and its relationship to the global, structuring struggles described earlier. I view these enduring struggles as among the sources of "discursive resources" for local contentious practice that partially structures more enduring struggles. Specifically, I focus on the idioms, vocabularies for making claims, and "modes of subjectification" that actors in the choice debate mobilized, some of which were appropriated directly from the global struggles described previously, and others of which had sources in other salient cultural structures. "Modes of subjectification" refers to the ways that language constructs individuals as particular sorts of persons endowed with particular needs, rights, and capacities (Fraser 1989:164–165).

The Language of Needs

Both sides in this debate framed their arguments in terms of the language of needs, using an individualistic psychological vocabulary to instantiate claims in this idiom. On the one hand, critics of the choice proposal claimed that choice would undermine the district's middle school philosophy, which was dedicated to meeting the "unique developmental needs" of young adolescents.[7] This feeling is illustrated in the following comments in a letter to the school board from a local teacher:

> We know that children must feel safe and included in order to focus on learning. . . . We fear for the educational, social and emotional health of all of our students in a "tracked" system, be it explicit or *de facto*. Young adolescents are particularly sensitive to issues of inclusion versus "special

treatment." The unique needs of ten to fourteen-year-olds guide middle school teachers and administrators as we continue to assess and refine our program.

Likewise, a letter from the Pro-Balance coalition to its sympathizers argued,

> Some of [our middle school teachers] are stars and some are not, but all of them use a variety of teaching methods, depending on the topic and the students they have. . . . We therefore should find it troubling that teachers in this "team of choice" would be asked to subscribe to the notion that "the parent is the customer" and to devote most class time to a single teaching method, regardless of the needs of their students.

On the other hand, members of the Parent Committee also invoked the "needs" of students as justification for their proposed program. For example, in their proposal to the school board, the Parent Committee suggested that reform instruction could continue to be offered in the district, but that parents of children who "learn better with direct instruction" should have that option as well. They were not seeking to impose their will on the whole district, the proposal explained, but only to be given permission to choose what was right for their own children. "At the middle-school level," the proposal argued, "few classrooms such as we have described [direct instruction classrooms] currently exist. Moreover, we feel that many children, certainly our children, learn best and feel most comfortable in this type of classroom." Likewise, following the school board's later decision to pilot the choice program, a representative of the Parent Committee stated in an interview with a local reporter that this decision signaled an acknowledgment by the district "that, just as there are different learning abilities and needs, there are different . . . ways that students learn. . . . All parents and teachers and children deserve and need choices and opportunities in education."

Local struggles "are interpreted through cultural forms that simplify, conceal, suppress, and give salience and priority to some ways of comprehending and participating in ongoing practice, in terms of some relevant subjectivities but not others" (Holland and Lave 2001:24). For example, how needs are constructed and interpreted within local debate, and the politics surrounding these processes, often conceal the broader processes of cultural struggle and change of which such local contests partake (Fraser 1989:164–165; Hall 1999:138). This has consequences both for local struggles and potentially also for more enduring conflicts because the meanings that are given salience in local debate make some actions appear more "reasonable" than others, and direct attention away from other potential inter-pretations and possible courses of action. In the present case, "needs talk" (Fraser 1989:164–165) served to partially mask the enduring struggles de-scribed previously, suppressing debate over questions of governance and power by displacing them onto other issues.

Specifically, the unquestioned status of children's "needs" as grounds for decisions in education "crowded out" (Holland and Lave 2001:25) potential struggle in the name of broader social concerns because it took for granted the individual rather than the collective purposes of public schooling. The way in which contestants on both sides of the debate mobilized psychologi-cal language presumed children's needs, rather than society's, as the taken-for-granted rationale for decisions concerning curricula and instruction. This common assumption gave tacit support to neoliberal models of educa-tion because it implicitly endorsed a conception of education as a private good. How best to address children's "needs" became the primary object of struggle, rather than the different conceptions of the purpose of public schooling implied by different arguments. However, if contestants had em-ployed language that brought broader social questions more effectively to the fore, then questions about the purpose of public education might have become a more direct object of local debate, with possible translocal effects.

The Language of Equal Treatment

Opponents of the choice proposal contended that the notion that the "par-ent is the customer" enables a "vocal minority" to command "special treat-ment" for select groups of parents and their children, at the expense of the greater good. For example, one parent warned the board,

> It is a very dangerous trend to allow groups of parents to set their children apart from the whole and to allow "private schools" within the public school system. Our school has the responsibility toward our community, toward our students, to protect and oversee the quality of education its public receives.

Again invoking the values of equal treatment and the common good, an-other critic characterized the proposal in a letter to the board as "a petition asking you to provide a separate program for 'their' children, a program where parents would be given the power to dictate particular teaching meth-ods and materials," warning, "if you approve their request, you will open the door for dozens of splinter groups who will demand that you respond to their pet programs." In a similar vein, members of the Pro-Balance coalition argued that the fulfillment of the Parent Committee's demands would draw scarce resources away from meeting the needs of the broader community. For example, a parent member of the Pro-Balance steering committee wrote to the board,

> To become deeply involved with every request for preferential treatment from concerned parents will seriously impact the quality of decisions made regarding the future course of the entire district. The board cannot afford to become distracted by this group of activists and must simply "just say no" to their request and move on to serve the greater needs of the district.

These examples illustrate a contrast between the individualistic discourse of the Parent Committee and that of the Pro-Balance group. The latter used the idiom of the common good, instantiated through a vocabulary of expert administration. This language contrasts the laissez faire, neoliberal vision of government implied by the choice proposal with a vision of activist government in which the state, via the school system, is obliged to "protect" the public by ensuring that all students receive the same "quality" of education.

However, the common-good arguments utilized by Pro-Balance generally implied a procedural rather than a substantive conception of equality, limiting the role of the state to the assurance of equal treatment. This mode of representation also suppressed explicit discussion of the deeper ideological issues related to competing models of educational governance at stake in the debate—for example, whether the schools should promote values that may differ from those of parents and whether parents have the right to opt out of programs determined by the state to be important for developing particular values and dispositions in its citizens.[8]

The Language of Partnership versus the Language of Rights

Other discourse strategies more directly engaged translocal struggles. For example, opponents of the choice proposal argued that parents should be "partners" with teachers and schools, instantiating these claims through a vocabulary of professional expertise that directly appropriated the terms of enduring struggles over educational authority. Using this vocabulary, critics of the proposal argued that parents should "respect" teachers as "dedicated professionals" and "support" them to help provide an excellent education for all children. For instance, a letter to the school board by a local teacher challenged the Parent Committee's market discourse by arguing that this mode of representing schooling endows parents with illegitimate authority over the work of teachers:

> The choice isn't about math or teaching technique but about the model of education we want in this district. On one hand, the model now in place in [Shady Hills] is one in which *education is viewed as a complex process, where teachers are accorded a degree of professionalism,* and where parents are viewed as partners working towards the same goal as the staff—quality education for all our children. Contrast this with a model growing out of a view of parents as "customers" of the district. Education then can't help but to be viewed as a "business." And, as is true in advertising a business, "catch phrases" abound. Suddenly the complex process of teaching and learning is reduced to unidimensional one-liners such as "new, new math," "direct instruction," or "discovery learning." While these phrases may capture a part of the picture they are woefully inadequate to convey the richness of what actually occurs in a classroom or school setting. In recent years in the district there seems to be an unsettling trend for parents to assume this "customer" role and from that vantage point to believe they have a right to "order" from the educational "menu" those things they decide, as customers, they desire. This can be a particular textbook, a particular narrowly defined teaching style, or a particular learning environment. *This relegates the professional educator to the role of "short order cook" to fill the orders.* Perhaps this model works in a fast food restaurant, but not in a math class, or school, or district. [emphasis added]

Other critics of the choice proposal made similar protests, arguing that the "parent as the customer" was an inaccurate, disrespectful, and inappropriate notion with dangerous implications. For example, a letter to the editor of the local paper from the husband of a teacher argued, "[A]ll of this recent school controversy . . . amount[s] to disrespect for school teachers." Another critic of the choice proposal argued in a letter to members of the school board, "Teachers are not 'robots' that can be 'programmed' to carry out the latest whim of 'customers.'"

This struggle over issues of professional expertise played out primarily through conflict over how to properly characterize the relationship between parents and schools, that is, over the politics of identity. An exchange that took place on the pages of the local weekly newspaper in early February illustrates this. The exchange was prompted by the publication of a letter to the editor from a parent and prominent community member in which the writer, using the idiom of partnership, challenged the validity of the "customer" argument on both economic and moral grounds:

The concept of parent as customer over-inflates the amount of tax dollars each parent gives to the school, and worse, puts the schools and parents in adversarial roles, where, in fact, both need to be partners, working together for the education of all children. The state, and in our town, the local tax-payers, are the customers, which is to say, they pay for our schools, which is why they are given the task of electing members of the Board of Education, who then provide, with the state, an educational philosophy the community desires. . . . The School Board need to listen, support and respect our principals and teachers as professionals who are capable of making knowledgeable decisions about curriculum and teaching methods.

Its publication prompted the following challenges in letters from advocates of the choice plan, also printed in the local weekly:

I find it strange that [the writer of the original letter] considers treating people as customers "adversarial." I hope she does not treat her doctor—or lawyer, or shop clerk—as an adversary. I do believe though, that she still expects them to treat her as a customer. I have heard this disingenuous "adversary" argument in the past as a reason to avoid competition in the schools. I guess that now we see it also as a reason to convince parents not to expect any kind of service. . . . I hope her letter will not take us back a few years, when the discourse was often about the "disrespectful" way parents may or may not talk to teachers, rather than on what they have to say about education.

One wonders how a citizen . . . in the name of "working together," can demonize and publicly vilify people whose ideas differ from hers. . . . One wonders about the true motivations of a parent Site Council and district committee member . . . who says that parent involvement means that parents should be restricted to passively endorsing the decisions of teachers and administrators, and that the Board of Education should surrender to the tyranny of inflexible "professionals" who are their employees. . . . One wonders about the understanding of a former PTA Council officer . . . who would deny to public school parents the right to make judgments about the educational approaches that work best for their own children (and no one else's).

These two letters position parents as rights-bearing citizens with the capacity and obligation to act in defense of those rights and of their children. Moreover, by means of the idiom of consumer rights, their arguments frame the conflict itself as a struggle for freedom and self-determination. Calling attention to the statuses of "citizen" and "district committee representative," the second letter moves outside the discursive frame of "partnership" in order to critique it from the standpoint of equality, casting advocates of partnership as antidemocratic.[9] Likewise, the letter also implies that the "professional" status accorded to teachers in the idiom of partnership is not only a smoke screen for concealing undemocratic practices (and, perhaps, also teacher incompetence) and coercing parents, but also a perversion of the appropriate authority structure. It reverses the hierarchy implicit in the language of partnership by positioning teachers as providers of a service for which parents are paying consumers.

These examples illustrate how contestants in the debate appropriated themes from long-standing struggles over both professional expertise and models of government. While the Pro-Balance coalition drew its arguments from the language of professionalism (and, to a much lesser degree, social democracy), the Parent Committee used neoliberal language to challenge this discourse by invoking the notion of consumer rights. This latter frame relies upon a specifically neoliberal conception of equality as a matter "of guaranteeing *individual choice* under the conditions of a 'free market'" (Apple 1989:35). In education, this mode of constructing equality has helped create a set of interpretations that gives meaning to the frustrations of parents who disagree with designated education experts, articulating their feelings of discontent in a coherent narrative that blames their troubles on a "monopolistic" system that denies parents their "rights" as "consumers" of the schools.[10] By contrast, the discourse of Pro-Balance implied a view of equality more consistent with social democratic models, which accord government responsibility for regulating the conditions under which individuals pursue their self-interests (based on a view of individuals as situated in structured relations of inequality that government can help equalize).

THE "TRAFFIC" BETWEEN LOCAL
AND GLOBAL STRUGGLES

> In the course of local struggles, marginalized groups create their own practices. Participants in these groups both are identified by these practices and often identify themselves as "owners" of them. These practices thus provide the means by which subjectivities in the margins of power thicken and become more developed and so more determinant in shaping local struggles. (Holland and Lave 2001:19)

Individuals and social groups are not just subject to globally powerful structures "but [are] also and critically appropriators of cultural artifacts" (Holland, Skinner, Lachiotte, and Cain 1998:17). Such acts of cultural appropriation are always structured by the conditions of (and particularly the resources available in) particular contexts. They are "improvisations, from a cultural base and in response to the subject positions offered *in situ*" (1998:18). For example, attempts to authorize behavior through the cultural production of identities (such as parents seen as "customers" of schools) can be understood as responses to the situations in which individuals find themselves, and to the modes in which they are positioned and addressed. The following account of the process by which the Shady Hills middle schools were created (from an interview with a parent who came out in favor of the choice proposal) illustrates this dynamic of structured improvisation, suggesting how the customer identity might have emerged as a response to how parents were positioned by the discourse of partnership:

> When my four children entered the public schools, I became a good active parent. . . . I was appointed to the . . . committee [charged with devising a new plan for the middle schools] and we . . . looked at what we were told was all the current research on education. Our new Superintendent . . . believed [in] the new educational reform movement, which included new research on the way the brain works, . . . where students need to be in a place where they are known and appreciated, and that knowledge needs to be conveyed in a way that connects [with their experience]. . . . We were given a lot of research . . . and told, "This is it. There is no other research." . . . We accepted all this printed stuff. . . . So this proposal was written recommending a new kind of middle school. . . . The . . . PTA opposed the change. . . . The teachers, district, and principals told us that "We the professionals" will now implement the change. The message was very clear that . . . those of us who are parents are just parents after all. . . . I felt I'd been used on that committee. . . . When they opened the new schools the emphasis was more on pedagogy and self-esteem. . . . They did away with tracking and laning, but they lowered the standards. . . . I still believe that parents are the consumers of education.

This narrative alludes to the roles parents may occupy within the discourse of partnership: One is only a "good, active parent" as long as one defers to the "professionals" in the end. Or, as this parent seems to be doing, one can struggle to find alternative idioms that endow parents with greater capacity and authority. The notion that parents are the "customers" of schools offers parents not only a set of solutions, but also (and perhaps more importantly) a way of talking about schools that addresses some of their individual frustrations and interests in ways that the normative idioms of "parent involvement" and "partnership" do not. Resisting the modes of subjectification entailed by the latter, the Parent Committee and their allies sought a new way of articulating the relationship between parents and schools, borrowing a vocabulary from neoliberal critics that endowed them with greater authority.

The Pro-Balance coalition reacted to this subversive move by challenging the legitimacy of the Parent Committee's economic idiom, asserting the sanctity of equal treatment and the "common good" as the organizing principles for public education and insisting on "partnership" as the role consistent with those principles. As one letter from a Pro-Balance ally implored, "Parents concerned about the quality of their child's education can find many positive ways to support our dedicated teachers and become partners in creating the best educational environment for ALL children." The attempts to reposition parents in the "booster" status implied by the language of partnership is perhaps most clear in a statement from a letter by the Pro-Balance steering committee, written in response to HOLD's original proposal for the creation of "middle school math houses" (which later became the proposal for "teams of choice"): "Despite the climate of distrust being fostered by HOLD activists, we believe that most parents respect the difference between appropriate and inappropriate parental input on curriculum issues." Similarly, a letter from a member of the Pro-Balance steering committee to the school board warned:

> The common perception is that the board is reconsidering HOLD's demands simply to "get them off our back." If this attitude affects major policy decisions, school governance in [Shady Hills] becomes a whole new game. Hundreds of parents who steadily support the schools in quiet and positive ways will either give up or adapt to destructive tactics. Scores of teachers who have earned our respect as professionals will face an increasingly hostile political climate (yes, it can get worse). And, ultimately, our children will get an education guided not by wise decisions but by frantic and uninformed voices.

My analysis suggests that individuals on either side of the choice question were not necessarily moved by any prior commitment to a particular model of government, but instead utilized the idioms at their disposal to accomplish pragmatic goals. For example, I interpret the way in which the Parent Committee appropriated neoliberal discourse as a strategic response to the alternatives presented them, rather than an expression of prior commitment to neoliberal ideology (cf. Apple 1996:42–67). Members and allies of the Parent Committee were not motivated to reconstitute the meaning of public schooling, but to secure a particular kind of instruction for their children. They utilized market idioms to argue their case because other methods had been unsuccessful. That parents who proposed the "Collaborative

Learning Team" were willing to piggyback on the Parent Committee's petition for direct instruction teams in order to secure a preferred instructional
environment for their children attests to the contingent, improvised nature
of participation in local contentious practice, since the two groups would
typically find themselves on opposite sides of struggle over curriculum.
Moreover, the way in which a local debate suppressed discussion of broader
ideological issues may also have contributed to the outcome of the dispute;
for example, more explicit discussion of the implications of the choice proposal for tacit conceptions of schooling (for instance, as a public or private
good) might have prompted different kinds of action.

With respect to struggle over professional expertise, it is less clear whether
individuals on either side acted out of prior ideological commitment. Nevertheless, what is significant for this case is how they made pragmatic use of the
idioms at their disposal to challenge their opponents' arguments. I have already explained why the customer identity was appealing to members of the
Parent Committee. What might seem less clear is why parent members of Pro-
Balance would use a discourse of professional expertise that relegates parents
to a "supporting" role in decisions concerning their children's education. My
analysis is that parents involved in Pro-Balance were not necessarily more disposed than their opponents to defer blindly to teachers' authority. Rather, they
were satisfied with the district's adopted math program and perhaps also more
committed (if in a somewhat contradictory way) to a common-good model of
education, and used the "teacher as expert" discourse as a strategic means to
defend the existing system and critique their opponents.

Motivation aside, as forms of constitutive social activity, these pragmatic choices and their effects on local events have significant structural implications, both locally and perhaps also more widely, in a number of specific
ways.[11] For example, *processes* of struggling such as those described in this
paper promote the development and "thickening" of identities (for instance,
parents as customers of schools), which, over time, can exercise increasing
influence over action, both locally and potentially elsewhere as well (Holland and Lave 2001:19). In the present case, the process of debate also reinforced neoliberal understandings of education by leaving assumptions about
the private purposes of public schooling unexamined.

Likewise, the *outcome* of such conflicts also has structural implications.
Such victories promote the simultaneous creation of both new institutional
mechanisms and new norms, conventions, and assumptions that, if invoked
with enough frequency and success, can produce incremental shifts in the relative institutional and cultural force of competing models of educational authority and governance, both locally and translocally. The Shady Hills school
district, in its handling of the crisis, abdicated responsibility for promoting any

substantive values beyond that of choice itself. The resolution of the conflict not only suppressed debate over these broader issues, but also lent tacit support to the conception of education as a private rather than a public good. From a structural standpoint, the Parent Committee's victory clearly dealt a local institutional blow to both professionalization and the common-good view of schooling, elevating the authority of parents over that of teachers in relation to judgments about curriculum and instruction and endorsing a neoliberal model of education governance. In the years since this debate, the choice option has indeed become institutionalized in Shady Hills. What began as a pilot program restricted to the sixth grade has since become a permanent policy, and has expanded to include other grades in the middle schools.

Successes such as the Parent Committee's not only establish local legitimacy and precedent for new modes of representing parents and schools, but also help promote the use of these idioms elsewhere, especially when publicized beyond the local site (as this case was). For example, parents in other parts of California and the nation have also (and with varying success) lobbied for choice as a reaction against changes in mathematics education (see http://mathematicallycorrect.com/). The direction of influence among these events is not always clear. What is certain, however, is that well-organized groups in this and other educational debates actively share stories, publicize their efforts, and imitate each others' strategies.

Of course, the longer-term consequences of this and similar local struggles remain uncertain in the present; structural change, as contrasted with mere "ripples in the surface" of enduring structures, can be more clearly grasped in retrospect. The point of this chapter has been to illustrate the open-ended, contingent, and indeterminate processes that mediate between local and global struggles. The outcome of the choice debate was not the inevitable effect of structuring global processes driven from the top down, but the contingent result of open-ended local practices of cultural production that clearly partook of translocal struggles, but did so in complex ways shaped by the exigencies of local conflict. Had local events transpired differently (for example, had parents adopted different discourse strategies, or had they been positioned differently by their opponents), the conflict might have seen a different resolution.

ACKNOWLEDGMENTS

I want to thank the parents and educators who made this research possible by sharing their experiences along with their personal archives of articles and correspondence related to the debate; Hugh Mehan and F. G. Bailey for guidance and support throughout the research and original writing of this chapter; Mira Katz, Holly Hart,

Nicole Holland, Raquel Farmer-Hinton, and my colleagues in the 2000–2001 Spencer Foundation Advanced Studies Institute, "Reconsidering the Interrelationship of Anthropology and Education," for insightful comments on earlier versions of this manuscript; and Katie Anderson-Levitt for helping me bring it to publication. The larger project from which the data for this chapter are drawn was funded by a Doctoral Research Fellowship from the American Educational Research Association and the Spencer Foundation in 1996–1997, and a Dissertation Fellowship from the Spencer Foundation in 1997–1998.

NOTES

1. For a concise overview of cross-national similarities in recent education reform, and their relation to broader processes of cultural, economic, and political change, see Burbules and Torres 2000:1–26. For a useful comparison of how market reforms have played out in various national contexts, and a discussion of how their similarities have generally been conceptualized within the sociology of education reform, see Whitty, Power, and Halpin 1998:31–47.

2. A pseudonym.

3. That the district is relatively small and that the contestants in this debate comprised a vocal, elite minority of the total parent population (even while both competing groups claim to speak for parents in general and one group to represent the interests of the district at large) does not limit their significance as objects of analysis. Indeed, debates in education that produce profound effects are frequently precipitated by the work of a tenacious few agitators (who often strive and sometimes manage to create the impression of representing larger constituencies). Moreover, the case described here is far from unique; a number of recent studies document similar cases of parents organizing to oppose progressive educational reforms, using a range of theoretical perspectives (cf. Apple 1996: 42–67; Graue and Smith 1996; Miller-Kahn and Smith 2001; Rosen 2001; Wells and Serna 1996).

4. Among these conditions were: that decisions concerning alternative programs must be considered in the context of both budgetary constraints and the willingness of teachers to participate in such a program; that "sufficient demand" among parents must be demonstrated before such a program could be created; that children in alternative programs will be required to meet the district's academic expectations; that parents of students in alternative programs might be required to pay additional costs associated with their child's participation; and that space would be set aside in each classroom for children with special needs.

5. For authoritative discussions of these intellectual traditions, and their relative power over practices in U.S. schools and schools of education, see Cremin 1964 and Lagemann 2000.

6. Lagemann (1993) offers a historical discussion of why the cultural transmission model has for so long dominated popular conceptions of the profession.

7. Beginning in 1988, parents participated actively in a lengthy process of community discussion, that culminated in the transformation of the junior high schools into middle schools in 1991. This transformation was accompanied by the adoption of a middle school philosophy based on the claims of educational research concerning the particular developmental needs of young adolescents.

8. In contrast to this general tendency, an email from a parent to the new superintendent (who assumed his responsibilities in the midst of the crisis) did hint at this broader struggle over models of government and the role of schools in producing particular kinds of citizens: "I do not consider myself to be in a consumer relationship with the [district]. The school owes it to all members of our community and the larger world to produce responsible young adults who know how to get along in diverse situations."

9. The letter also evokes familiar cultural figures that appeal to common anxieties about modern bureaucratic and political culture: the "false expert," the "puppet democracy," the oppressive "system" and the "underdog/outsider" who takes them all on in the name of justice and right.

10. For a complementary analysis of the influence of neoliberalism on local educational policymakers and administrators in California (which suggests an explanation for why the Parent Committee's language may have resonated with members of the Shady Hills school board), see Wells, Slayton, and Scott 2002.

11. My thinking on problems of structural change is informed by Bailey's theory of "bridge actions" (1960:248–269), as well as his analyses of politics more generally, and particularly of political change (for example 2001a, 2001b).

REFERENCES

Apple, M. 1989. The politics of common sense: Schooling, populism, and the new right. In Critical pedagogy, the state, and cultural struggle, edited by H. Giroux and P. McLaren. Albany: State University of New York Press.

———. 1996. Cultural politics and education. New York: Teachers College Press.

Bailey, F. G. 1960. Tribe, caste, and nation. Manchester: University Press.

———. 2001a. (1969). Stratagems and spoils: A social anthropology of politics. Boulder, CO: Westview Press.

———. 2001b. Treason, stratagems and spoils: How leaders make practical use of beliefs and values. Boulder, CO: Westview Press.

Burbules, N., and C. Torres, eds. 2000. Globalization and education: Critical perspectives. New York: Routledge.

California Department of Education. 1992. Mathematics framework for California public schools: Kindergarten through grade twelve.

Cremin, L. 1964. *The transformation of the school: Progressivism in American Education, 1876–1957.* New York: Alfred A. Knopf.

Fraser, N. 1989. *Unruly practices.* Minneapolis: University of Minnesota Press.

Graue, M. E., and S. Z. Smith. 1996. Parents and mathematics education reform: Voicing the authority of assessment. *Urban Education* 30 (4): 395–421.

Hall, K. 1999. Understanding educational processes in an era of globalization: The view from anthropology and cultural studies. In *Issues in education research: Problems and possibilities,* edited by E. Lagemann and L. Shulman. San Francisco: Jossey-Bass Publishers.

Holland, D., and J. Lave, eds. 2001. *History in person: Enduring struggles, contentious practice, intimate identities.* Santa Fe: School of American Research Press.

Holland, D., D. Skinner, W. Lachiotte, Jr., and C. Cain. 1998. *Identity and agency in cultural worlds.* Cambridge: Harvard University Press.

Labaree, D. F. 1997. *How to succeed in school without really learning: The credentials race in American education.* New Haven, CT: Yale University Press.

Lagemann, E. 1993. Reinventing the teacher's role. *Teachers College Record* 95 (1): 1–7.

———. 2000. *An elusive science: The troubling history of education research.* Chicago: University of Chicago Press.

Lingard, B. 2000. It is and it isn't: Vernacular globalization, educational policy, and restructuring. In *Globalization and education: Critical perspectives,* edited by N. Burbules and C. Torres. New York: Routledge.

Luke, A., and C. Luke. 2000. A situated perspective on cultural globalization. In *Globalization and education: Critical perspectives,* edited by N. Burbules and C. Torres. New York: Routledge.

Miller-Kahn, L., and M. L. Smith. 2001. School choice policies in the political spectacle. *Education Policy Analysis Archives* 9 (50). Available at http://epaa.asu.edu/epaa/v9n50.html. [Accessed December 6, 2001.]

Rosen, L. 2001. Myth making and moral order in a debate on mathematics education policy. In *Policy as practice: Toward a comparative sociocultural analysis of educational policy,* edited by M. Sutton and B. A. U. Levinson. Westport, CT: Ablex Publishing.

Shore, C., and S. Wright, eds. 1997. *Anthropology of policy: Critical perspectives on governance and power.* New York: Routledge.

Wells, A. S., and I. Serna. 1996. The politics of culture: Understanding local political resistance to detracking in racially mixed schools. *Harvard Educational Review* 66 (1): 93–118.

Wells, A. S., J. Slayton, and J. Scott. 2002. Defining democracy in the neoliberal age: Charter school reform and educational consumption. *American Educational Research Journal* 39 (2): 337–361.

Whitty, G., S. Power, and D. Halpin. 1998. *Devolution and choice in education: The school, the state and the market.* Philadelphia: Open University Press.

Zeichner, K. 1991. Contradictions and tensions in the professionalization of teaching and the democratization of schools. *Teachers College Record* (3): 363–379.

CHAPTER EIGHT

WORLD CULTURE OR TRANSNATIONAL PROJECT?

Competing Educational Projects in Brazil

Lesley Bartlett

In 1963, after participating in a series of remarkably successful literacy programs with peasants in Brazil's Northeast, Paulo Freire was invited by the populist president João Goulart to head a national commission on popular culture. From that position, nestled within the national bureaucracy and using the state's machinery (or creating new machinery when necessary), Freire was expected to diffuse his radical political-educational philosophy and his literacy pedagogy (Beisiegel 1974; Fernandes and Terra 1994). Before he could begin, however, the military staged a coup: They unseated Goulart, imprisoned and eventually exiled Freire and many of his colleagues, and made radical literacy work illegal. The accession to the state apparatus of a politicized education dedicated to the redistribution of social power was nullified before it could even begin to make, much less fulfill, promises.

Nevertheless, in Brazil, as in much of Latin America, two distinct educational projects continue to compete for the privilege of defining adult literacy and delimiting its uses. The first, what I call economic efficiency, places schools in the service of economic goals of the market and transnational capital, the state, and/or the individual student. This educational project predominates in systems of formal schooling. The second, which is widely known as popular education, advocates a more communal vision of human development, wherein schooling directed at marginalized populations encourages political action and aims to redistribute power and wealth (or at least criticizes the lack of equity and promotes social, rather than individual, explanations for that lack). This educational project predominates in nongovernmental organizations (NGOs).

In this chapter, I use the case of Brazilian adult literacy to reconsider neo-institutionalist theory, a.k.a. world culture theory. World culture theory illuminates the important role of national actors who sponsor cogent, universalist models of schooling, including organization, pedagogy, and curriculum (Boli and Ramirez 1992; Meyer and Ramirez 2000; Meyer et al. 1997). However, I contend that when world culture theorists reduce educational variety and debate to a singular "model" for schooling, they do not adequately address the different interest groups, organizations, philosophies, and pedagogies involved in the elaboration and promotion of educational projects. When they ignore nonformal education, they overemphasize the state as actor and eliminate from consideration important educational programs (some of which increasingly engage directly with government-sponsored programs). When world culture theorists attend to reforms generated only in the West (or in international organizations and/or educational professions dominated by the West) and "diffused" to other places, they downplay the possible impact of other projects. And when they fail to detail the power relations and the logic behind the importation and (partial) implementation of educational reforms, they grant too much autonomy to organizations while de-emphasizing human agency.

In the case of Brazil, I argue that nonformal, popular educators, inspired by Paulo Freire, mounted and continue to maintain a significant politicized educational project that was embraced within Brazil and across the globe by nonstate actors, such as teachers, professors of education, and NGOs. These actors used (and continue to use) it to contest the dominant, widespread, economistic educational project—that is, precisely the currently hegemonic (but not permanent, ubiquitous, or omnipotent) world model of mass schooling described by world culture theorists.

In order to make this argument, I first provide a brief overview of world culture theory. I then introduce the concept of an educational project and explain why I think it is a richer concept than "models" or "world culture," with much more to offer to studies of education. Finally, I discuss the historical development of Brazilian literacy projects, using the case to elucidate my appreciation and my criticisms of world culture theory.

WORLD CULTURE THEORY: A BRIEF SUMMARY

Reacting to overly functionalist theories of education, which claimed either that mass education laid the foundation for political and economic development or that mass education was a tool used by elites to maintain social control, the sociologically minded neo-institutional theorists (including John Meyer, Francisco Ramirez, George Thomas, John Boli, and others) developed a new explanation for the striking similarity among national systems of mass education (Boli and Ramirez 1992; Meyer and Ramirez 2000; Meyer et al. 1997). They argue that two revolutions in human thinking—the shift from corporatist or localist views to the nation-state, and the shift from God-centered to individual-centered thinking—enabled and indeed encouraged young nation-states to implement the model of mass schooling that originated in the West. According to the theory, this model then "diffused" over the nineteenth and twentieth centuries from the "center" to the "periphery," as the elite of young nation-states, eager to declare their modernity, adopted the model. World culture theorists argue that systems of education "are more homogeneous across countries than would be predicted by the actual variability of national societies and cultures" (Meyer and Ramirez 2000:119). Over time, they contend, the "impact of particular endogenous national political, social, and economic characteristics on national educational systems declines" while the influence of international organizations and the increasingly standardized, professionalized field of education intensifies (2000:119).

The macro-sociological, neo-institutionalist approach helpfully reminds us of the remarkable similarity in mass schooling, despite significant differences among nation-states in terms of their economic, political, and social arrangements. It helps to unseat erroneous, simplistic causal arguments that claimed, for example, that the expansion of mass education led directly to democratization, national industrial revolution, or other forms of capitalist integration. Finally, the approach suggests the importance of imagined communities by emphasizing the local elite's symbolic use of mass schooling to declare their status as a "modern" nation-state.

However, I argue that world culture theory has several weaknesses that an anthropological perspective illuminates. First, the theory overemphasizes the power of elites. On occasion, the writings of world culture theorists grant agency to the organization itself, as when they suggest that organizations tend toward isomorphism. This move obscures the agency of nonelite individuals and collectivities; it confuses official policies with actual practices (Sutton and Levinson 2001).

Second, world culture theorists have not paid sufficient attention to the factors that persuade states to adopt the dominant model of schooling. At times, they suggest that national-level elites, in their yearning to be modern, embrace and unproblematically emulate these models as symbols of "world culture." While world culture theorists' later writings recognize the importance of world-system integration and the role of powerful international organizations such as the World Bank and (to a lesser extent) UNESCO, they don't address questions of power, imposition, and hegemony. Yet, as Gita Steiner-Khamsi and others have shown, nations "borrow," and often adapt, dominant reforms to legitimize (inter)national or local positions, rather than simply conform to or embrace international norms (2000). Anthropologists have long rejected the concept of diffusion, insisting on a closer on-the-ground study of the complex reasons for which some people adopt and others resist certain processes.

World culture theory posits that first Western nation-states and then international organizations and knowledge regimes (such as educational professions) diffuse the model of schooling from the center to the periphery (Meyer and Ramirez 2000). This approach discounts the influential models developed in and communicated from the South to other parts of the globe. Popular education, and in particular Freirean-style critical pedagogy, is a prime example. Indeed, by examining only formal schooling, world culture theory excludes a multitude of nonformal, noninstitutionalized educational efforts from consideration. However, in some regions (for example, Latin America, India) and with some populations (for example, the poor, ethnic and/or linguistic minorities, women, people living in rural areas), nonformal education historically provided a main source of schooling (see, for example, Hausmann 1998; Kindervatter 1979; La Belle 1986; Oxenham 1975; Sheffield and Diejomaoh 1972; Singh 1987; Srinivasan 1977). In the 1970s, deschooling enthusiasts invested significant energy in nonformal education. By ignoring nonformal schooling, world culture theorists also overlook an important phenomenon—the growing tendency of nonformal education programs to work with, and at times to be funded by, state or local governments (for example, UNESCO 1986; Yadav 1987). In fact, states increasingly govern through "partnerships" with NGOs and private entities (see Holland et al. 2002).

Fourth, world culture theorists discuss a single model of schooling, which they insist includes tendencies as diverse as the individual and national pursuit of economic mobility and an emphasis on human rights (Meyer and Ramirez 2000). This homogenization conceals the very real differences in ideologies and social actors involved in the struggles over schooling. It marginalizes competing notions of schooling and its uses. Instead of accepting a "big tent" model, I propose that we think of *educational projects,* or configurations of social actors, institutions, financial sources, discourses, philosophies, and pedagogies.

EDUCATIONAL PROJECTS

I derive the concept of an educational project from Omi and Winant's racial formation theory (1994). For Omi and Winant, racial formation involves both cultural and social structural processes. A racial project "is simultaneously an explanation of racial dynamics and an effort to reorganize the social structure along particular racial lines. Every project is necessarily both a discursive or cultural initiative, an attempt at racial signification and identity formation on the one hand; and a political initiative, an attempt at organization and redistribution on the other" (Winant 1994:24). I argue that we should consider educational projects in a similar vein. Educational projects, which include theory, pedagogy, philosophy, training, and institutions, plus a variety of social actors and social practices, also simultaneously work at the cultural/discursive and political/structural levels. They shape how we think about things as intimate as knowledge, intelligence, and personhood, even while they structure our material world, including the kinds of buildings we congregate in, the use or rejection of desks, blackboards, walls, or classrooms, the availability of instructional materials, and the content and form of books we read (or don't read).

I argue that there is rarely a single, coherent educational project, as claimed by world culture theory. Instead, projects intersect with subordinate, less cohesive ideas of education in their implementation. Further, multiple projects compete for hegemonic control of the public's imagination. In fact, educational projects are race and class projects, in that they benefit some groups and penalize others—though in general this ideological work is obscured, for example through the naturalization of cultural arbitraries such as "intelligence" or "worth."

The social actors involved with these projects strive for legitimacy within a social and discursive field (Bourdieu and Wacquant 1992:41), aiming to control the structures, experiences, and ends of schooling. Building on Gramsci's notion of the state as "terrain" for a war of position among people with conflicting interests, I imagine groups organized through and around educational projects; victory in the competition grants an ideological legitimacy within the "common sense" of the general population, structuring perception, feelings, and experience (Gramsci 1971; Williams 1977). Educational projects proffer discourses on the purpose of education. By selecting which courses to offer where, to whom, which pedagogy to employ, and what classroom dynamics to foster, project proponents structure educational experiences. Less evident, but just as important, is the work done in these projects to shape students' subjectivities.

The concept of educational projects, then, widens our consideration of the impact of the world system on mass schooling. People working in and through international organizations like the World Bank and national organizations such as ministries of education do, indeed, have extensive access to and influence over the institutions, funding, curricula, pedagogies, and methods of assessment involved in schooling. But their decisions are not uncontested. They too must face the challenges mounted by other interest groups, such as social movements and political organizations, which work through alternate organizational structures and employ competing discourses regarding the purposes of education and the meaning of intelligence, among other issues. In short, the dominant educational project must accommodate, permeate, or surmount rival educational projects to establish hegemony.

In comparison with the notion of a singular educational model, the more nuanced concept of an educational project better captures a sense of change over time, simultaneously omnidirectional and multilateral nation-state relations, change originating from non-Western and nonelite sources, the use of educational programs in contests of political legitimacy, and conflict between and within groups, institutions, and/or discourses. The concept of educational projects encourages us to look more carefully at the discursive and cultural dimensions of the debate, as well as the outright political facets. Describing efforts as educational projects, rather than as a singular model of schooling, enables us to see educational politics as cultural dialogues rather than unidirectional movements.

The notion of educational projects encourages us to compare ongoing educational debates, or what Holland and Lave call "enduring struggles," across space and time (2001; see also Rosen in this volume). According to Holland and Lave, local debates mutually partake of and (re)produce enduring social structures. For example, thanks to globalization processes, capitalist class formations and discourses may increasingly circulate across the globe; however, they are grappled with, engaged in, and remade in unique ways in different situations.

The same is true for educational debates. For example, the divide in Brazilian literacy politics between economistic and popular education approaches partakes of the same enduring struggle as the contest in the United States between schooling for democratic equality, social efficiency, and/or social mobility (Labaree 1997). According to Labaree, the history of schooling in the United States demonstrates a shifting among three goals: serving democracy by promoting equality and providing training for citizenship (democratic equality); serving society and the economy by training students to occupy positions in a stratified social order (social efficiency); and providing individual social advantage (social mobility) (1997). The balance between the goals changes over time, but the struggle remains. These enduring, transnational struggles, with their universalist discourses on education's purpose(s), manifest themselves in different ways across place and time. Our job as educational researchers lies in tracing the differences, as well as the similarities, in order to build theories about educational systems and their interrelationships.

CONTENDING LITERACY PROJECTS IN BRAZIL

In the remainder of the chapter, I discuss the history of adult literacy programs in Brazil in order to illuminate the advantages of the concept of an educational project and the limitations of world culture theory. In other venues, based on two years of ethnographic research in public and NGO classrooms in two Brazilian cities, I have examined the contemporary effects of contending educational projects in Brazilian literacy classrooms (see Bartlett 2001). Here, though ethnographic research informs my argument, I generally restrict my discussion to secondary sources concerning a critical period of literacy politics in Brazil, dating from the early 1960s.

Popular Education

Since the 1960s, two radically different literacy projects have vied for control of Brazilian educational resources and for influence over popular conceptions of education's purpose. The first, known throughout Latin America as "popular education," developed from radical Catholic liberation theology and leftist political organizing. Popular education places education in the service of politics, society, and culture—in short, in the service of human, rather than only economic, development. Proponents of popular education argue that schooling should be democratic, that is, open to all who desire to participate. They favor social over individual change, and encourage students to develop a social critique of the world. For example, they encourage students to see a rude or dismissive boss as a member of the exploitative owners of capital; they encourage women to place a personal experience of domestic violence in the context of patriarchy and poverty. For popular educators, literacy raises people's consciousness by fostering awareness of oppressive political and economic structures and providing the basic skills to intervene in them. Though popular education emerged from adult literacy politics (see below), it has expanded to encompass schooling for all ages and in all situations. It resembles what Labaree (1997) denoted as the democratic equality goal of schooling. However, unlike traditional efforts to promote democratic equality through schooling in the United States, popular education has since its inception emphasized culture as a critical arena for egalitarian struggles.

Popular education is epitomized by the 1960s experiments of educators like Paulo Freire in the Brazilian Northeast (see Beisiegel 1982; Brandão 1980; Fink and Arnove 1991; Freire 1970; Gadotti 1994; Kane 2001; La Belle 1986; Paiva 1973; Paiva 1980; Stromquist 1997; Torres 1995). In 1962, the young Northeasterner Paulo Freire rose to prominence for his radical humanist pedagogy developed to work with the poor of the region. His philosophy combined Christian notions of radical equality before God and a Marxist critique of unequal class relations. Freire excoriated the neocolonial relationship between "oppressor" and "oppressed," which alienated the oppressed from their true free selves. He believed that an egalitarian pedagogy would overturn class divisions and liberate both oppressed and oppressor.

Freire argued that the "banking model of education," in which the teacher owns knowledge and deposits it in the heads of students, contributes to oppression. Freire proposed teachers as the "revolutionary leadership" who should engage in "critical co-investiga[tion] in dialogue" with students (Freire 1970:68). Though not a social constructionist, Freire certainly was relativistic about truth and facts, and he valorized student experiences. His pedagogy centrally featured dialogue regarding generative words or themes determined through short-term experiential, almost ethnographic research with students in their communities. Theoretically, this dialogue should encourage "conscientization," or sociopolitical awareness, among students and teachers.

Thus, the theory of popular education views knowledge as embedded in and emerging from social context. It aims to conscientize the individual, and thus maintains a certain individualist orientation—yet it does so through social interaction, dialogue, and eventually social action.

In the early 1960s, popular literacy programs spread quickly. Literacy was a requirement to vote in the 1960s (indeed, it remained a prerequisite until 1988). Since 40 percent of the population at that time was illiterate, the Left saw the building of literacy as an opportunity to democratize the country and build a considerable populist political base. These organizations clustered in Brazil's poor and highly illiterate Northeast, where they were frequently sponsored by radical Catholic university groups and attached to emerging peasant labor unions (Lemos 1996). For example, the Movimento de Educação Brasileira (MEB), propelled by Catholic activist priests and parishioners and partially funded by government supporters, set up a network of radio schools throughout the Northeast (Costa, Jaccoud, and Costa 1986).

In 1963, after extensive involvement with cultural circles and educational initiatives throughout the Northeast, Freire initiated a literacy campaign in Angicos, Rio Grande do Norte. In an attempt to domesticate the suddenly radical turn of politics in the Brazilian Northeast, U.S. president Kennedy's Alliance for Progress partially funded Freire's literacy experiment. In Angicos, Freire trained young university students, the majority involved in the Catholic social action movement, to engage in "dialogical" literacy education. The story, widely reported throughout Brazil, is that within 40 hours of instruction in these classrooms, a sizeable group of *nordestinos* became literate. Then President João Goulart himself visited the isolated community to celebrate their achievement in the fortieth hour (Fernandes and Terra 1994). Soon afterward, Freire was invited by the populist national administration to implement his literacy method nationally. This plan was aborted by the military coup in 1964.

After the coup, military rulers arrested literacy activists, exiled Freire, and criminalized the conduct of Marxist humanist education. In its stead, the dictators substituted a vocationally oriented literacy program with clear ideological bases in human capital theory called MOBRAL (Movimento Brasileiro de Alfabetização, or Brazilian Literacy Movement) (Beisiegel 1974; Jannuzzi 1979; see next section). The military regime's crackdown on political activists fragmented the literacy movement, scaring some away from activism and sending others to the shelter of sympathetic Catholic communities, where they unobtrusively continued to do their work in isolation.

Ironically, because of Freire's exile, popular education spread internationally. For a decade, Freire worked at the World Council of Churches in Geneva. From that position, he actively participated in projects in Latin America and Africa, including Sao Tomé and Principé, Guinea-Bissau, and Chile. The Church's powerful and extensive networks broadcast his message. Freire also worked for a short time at Harvard University's School of Education, augmenting the interest of American college and university faculty members, who drew upon and thus publicized his work widely. Further, Freire's lively involvement in UNESCO enhanced global access to his ideas.[1] The internationalization of Freire's ideas marks an unusual reversal of the normal pattern of educational transfer from Europe and the United States to other areas.

With the gradual redemocratization of Brazilian society that began in the mid-1980s, many literacy activists resumed their political education work. Yet the movement never recovered its momentum (Beisiegel 1982). Literacy workers in the 1970s and 1980s remained adamantly nonformal, resisting incorporation into the state structure of schooling. Popular education survived with insecure funding, low teacher salaries, and high teacher and student turnover, in informal settings such as churches, worksites, union halls, and makeshift classrooms. The neoliberal political moment has eroded teachers' sense of belonging to a social movement and, to some degree, fractured popular education efforts.

Yet Brazilian popular education endures. Proponents organized in local, state, and national networks of popular education groups serve as aggressive and vocal watchdogs of federal, state, and local policy regarding adult education. For example, the Rede de Apoio à Ação Alfabetizadora do Brasil (RAAAB, the Brazilian Network to Support Literacy Action) holds regular regional and national literacy "fairs," where (in addition to providing teacher training workshops and forums for popular educators to analyze their literacy practices) advocates discuss literacy politics, issue communiqués, and plan strategies to affect policies. RAAAB also publishes a lively journal, *Alfabetização e Cidadania* [Literacy and Citizenship], dedicated to "the exchange and systematization of ideas inspired by the popular education paradigm and the implementation of government practices in that area." Further, several key organizations disseminate popular education on a national scale. The Instituto Paulo Freire and Ação Educativa [Educational Action], both in São Paulo, publish articles, books, and reports, maintain archives, provide consulting, and offer staff training for organizations across Brazil.

Further, popular education remains vibrant in Brazilian schools of education. There, scores of courses discuss the history of popular education, Freire's life, philosophy, and pedagogy, educational social movements, liberatory adult education, and the current state of popular education, while others incorporate Freirean approaches into their classroom practice and their teacher training. Further, many professors dedicate their research to understanding and expanding popular education (see, for example, Coraggio 1996; Costa 1998; Graciani 1997; Lovisolo 1990; Scocuglia 1997; Souza and Porto 2000). In these schools of education, (generally) middle-class students and professors enact equity concerns and opposition to neoliberal policies through their education work.

Contemporary nonformal literacy efforts continue in marginalized communities where public provision of education is weakest. For many of Brazil's poorest—migrants in large cities, subsistence farm families in rural areas, landless peasants, women in impoverished communities, and street children—popular education remains their only option, despite the considerable expansion of public schooling in recent years. In other areas, Freirean literacy programs serve as portals to the public system, wherein people learn emergent literacy skills (such as recognizing the alphabet or participating in class discussions) that embolden them to enroll in public school adult classes. These programs shape not only the students' but also the teachers' perceptions of education's purpose. Hundreds of working-class teachers with modest educational and income levels have been recruited to this particular vision of human development and social justice.

In the 1990s, NGOs began the slow and sometimes painful process of learning to collaborate with state and local government. In half of the states, literacy workers from NGOs, social movements, state and local government, and universities have formed "forums" to debate and propose literacy policies and initiatives. NGOs increasingly partner with governments. For example, First Lady Ruth Cardoso's renowned federal Programa Alfabetização Solidária [Solidarity Literacy Program], works in partnership with civil society and business enterprises. Two of the three NGOs I worked with in João Pessoa also took this path. One maintained a five-year contract with the Paraíba state government, through which the state government provided the financing but the NGO controlled the pedagogy. Another, in the wake of the decentralization of basic education, agreed to provide pedagogic consultancy to the local government as the latter increasingly assumed responsibility for adult education courses. The leaders of one neighborhood-based NGO bluntly informed me that they intended to "infiltrate" the local public schools with young teachers who had worked as catechists in the liberation theology–influenced Catholic base communities and had participated in Freirean teacher training. To this end, the NGO was providing small scholarships for their popular education teachers to attend university (Bartlett 2001).

These partnerships bring new challenges. Skeptics worry that governments are co-opting the critical voice of NGOs. And critics point out that governments have essentially hired NGOs to do their basic literacy instruction, but without affording the benefits, salary, or security that a public teaching position typically provides. In one of the NGOs with which I worked in João Pessoa, NGO teachers contracted by the state considered going on strike when they learned of the salary differential between themselves and public school teachers. Three years later, due to a political conflict, the state severed the partnership and disbanded the adult literacy program altogether. Thus, although the interchange between civil society and government literacy efforts has increased, possibly enabling NGOs to have more impact on the public system, the two educational projects remain distinct and at times antagonistic.

Economic Efficiency: Human Capital Theory and Beyond

Adapting Labaree (1997), I call the second educational project competing for the hearts and minds of Brazilians "economic efficiency." This paradigm

views education as serving economic, not human, development, assuming
that the former will lead to the latter. The project emphasizes individual economic mobility and national economic growth (although the presumed relationship between the two has shifted over time).

The long-standing economic efficiency approach to education was re-
juvenated by human capital theory, which emanated from U.S.-based economists in the 1960s and circulated widely through development agencies.
Human capital theory predicted that investment in the labor force's basic education would increase workers' productivity and improve efficiency, which
would theoretically promote economic expansion, provide more and better-
paying jobs for the poor, and produce greater profits for the owners of capital. Human capital theory provided an argument for government
expenditure of limited resources on education and training; thus, it was instrumental in promoting the twentieth century international expansion of
mass education that so interests world culture theorists. In Brazil, the spread
of human capital theory paralleled and, I posit, helped to propel the extension of basic education (see also Plank 1987; Plank 1996).

Human capital theory spread throughout the world quickly through
the auspices of international financial institutions such as the World Bank
and its regional subsidiaries, unilateral development lending institutions,
and other development organizations (Eshiwani 1989; Fonseca 1996; Jones
1992; Puiggros 1996). It appealed particularly to political entities heavily invested, both financially and symbolically, in state-directed, modernist capitalist economic development. Such was the case for the military dictatorship
that ruled in Brazil from 1964 to 1985. Between June 1964 and January
1968, the military dictatorship of Brazil signed 12 education accords with
USAID, each of them guided by human capital theory (Ghiraldelli
1994:169). The USAID funding that aimed to control educational reform
in the Northeast in the early 1960s found military leaders to be more re-
ceptive and ideologically aligned allies. By the end of the 60s, the World
Bank increasingly assumed the disbursement of aid to Brazil. Throughout
the 1970s and 1980s, the World Bank funded five major educational projects—two of which specifically addressed schooling in the Northeast (Fonseca 1996). In congruence with the dictatorship's interpretation of human
capital theory, several Bank projects concentrated on vocational education.

The economic efficiency approach to adult literacy came in the form of
the military government's massive MOBRAL literacy project (Bola 1984;
Cairns 1975; di Ricco 1979; Jannuzzi 1979; Nunes 1992; UNESCO 1975).
This literacy campaign, funded primarily by a national sports lottery, was in-
augurated in 1967 but saw its greatest expansion in the early 1970s; it was

terminated upon the redemocratization of the country in 1985. An estimated 2 million students and 4,500 functionaries were involved over the 18-year tenure of the program (Hollanda 1997). Across the nation, teachers were recruited through a variety of mechanisms, including television commercials urging (middle-class) patriots to "erase" illiteracy through classroom volunteerism. Some who had been active in popular education inadvertently slipped into the teaching corps. MOBRAL classes, often in nonformal settings, popped up across the nation.

The military government countered popular education's liberatory rhetoric with an economistic vision of literacy and education. The top-down campaign heralded the economic value of literacy, insisting that it would contribute to the capitalist development of the nation (Almeida et al. 1988). Instead of the critical "generative words" used by Freireans, such as *fome* (hunger), *moradía* (housing), or *saúde* (health), MOBRAL teachers began by teaching letters (vowels and then consonants) before "progressing" to syllables and finally words (Nunes 1992). The enduring struggle between social, whole-language approaches and economistic, phonics-and-skills approaches to language and literacy continues in Brazil and has materialized in many other places over the years, lately (and fiercely) in the United States.

Human capital approaches to literacy continue. The federal *Programa Alfabetização Solidária* (Solidarity Literacy Program), funded by the government as well as by UNESCO and national business enterprises, has in many ways superseded the failings of MOBRAL—for example, it works in partnership with NGOs, universities, and businesses, and as a decentralized institution is much more responsive to local needs. However, in his defense of the program in the *Folha de São Paulo* (31 October 2000), Minister of Education Paulo Renato justified the program as contributing to "better qualified labor and a more active citizenship." Human capital theory also remains popular in international development circles and the upper echelon of Brazilian education policymakers; both groups train in economic approaches to education, often at Western institutions (see Birdsall and Sabot 1996; Easton and Klees 1992; Harbison and Hanushek 1992). Further, the World Bank continues to have a significant effect on Brazil's educational policies, most recently in moves to shift funding from tertiary to primary schooling for children and in the form of a large Bank project conducted over a period of years in northeastern Brazil (Harbison and Hanushek 1992).

As an educational project, economic efficiency has evolved to fit neoliberal reforms sponsored by the development banks and partially executed by most Latin American governments since the 1980s (Henales and Edwards 2000; Levy 1986; Puiggros 1999). Proponents of economic efficiency have cooled on the idea of government intervention (Easton and Klees 1992). They increasingly recommend the marketization of education through increased user fees, the injection of market principles of competition and consumer choice through outright privatization of services, and increased standardization of curriculum and assessment through national testing. As world culture theorists predict, this approach increasingly influences the provision of education throughout the world. However, where it is implemented, how, and why depend largely on local conditions. For example, Chile's extensive voucher program has yet to be adopted in Brazil or other Latin American countries (Gauri 1998).

CONCLUSIONS AND IMPLICATIONS

As the preceding discussion illustrates, two educational projects vie for control of educational programs, pedagogies, and resources in Brazil. The first, popular education, predominates in nonformal education, though it occasionally infiltrates the public school system through a variety of means—by educating teachers or administrators in workshops or at university, by entering partnerships, or by forwarding students to the public system. Popular educators employ social approaches to language and literacy and insist that education can and should contribute to human development and social justice. The second, economic efficiency, prevails in the public schools. Proponents of economic efficiency studiously avoid the political aspects of schooling, utilize phonics-based instruction, and aver that education can and should contribute to economic development. The latter project, while dominant, is decidedly not the only model for education in Brazil.

The concept of educational projects allows us to understand adult literacy education in Brazil in a way that world culture theory does not. First, we perceive multiple, competing projects, rather than reducing great variety to a single model. Second, rather than privileging the state, the concept of educational projects allows us to consider the role of nonformal education and reveals the importance of nonstate institutions, such as the Catholic Church, as transnational actors. Third, rather than emphasizing educational innovations emanating from the North and West, it brings theories and activities from the South into view.

The concept of educational projects allows educational researchers to think about the enduring struggles over education's purpose(s), as expressed by diverse groups in disparate times and places. It allows us to examine the momentary, consecutive intersections of actors (international, national, state, and/or local), institutions, discourses, philosophies, pedagogies, and resources that constitute educational programs or reforms. While the notion of world culture indicates homogeneity, the idea of an educational project encompasses heterogeneity but also seeks to identify common elements and compare their roles over space and time. In this way, the concept furthers the work of a culturally grounded, comparative approach to educational research.

ACKNOWLEDGMENTS

I gratefully acknowledge the financial support of several institutions during this research project: Fulbright/IIE, the Inter-American Foundation, and the Social Science Research Council.

NOTES

1. This is not to deny that UNESCO also engaged in economic developmentalism, which it certainly did. UNESCO, like other organizations, has long been subject to competing visions of literacy (Jones 1990).

REFERENCES

Almeida, T. W. de, E. M. de Nascimento, and I. A. Veronese. 1988. História y situación de la evaluación en el Mobral. In *Enfoques y experiencias sobre evaluación de educación de adultos.* Michoacan, Mexico: CREFAL/UNESCO.

Bartlett, L. 2001. *Literacy shame and competing educational projects in contemporary Brazil.* Ph.D. dissertation, University of North Carolina at Chapel Hill.

Beisiegel, C. d. R. 1974. *Estado e educação popular.* São Paulo: Pioneira.

———. 1982. *Política e educação popular: A teoria e a prática de Paulo Freire no Brasil.* São Paulo: Editora Atica.

Birdsall, N., and R. H. Sabot. 1996. *Opportunity foregone: Education in Brazil.* Washington, D.C.: Inter-American Development Bank.

Bola, H. S. 1984. *Campaigning for literacy: Eight national experiences of the twentieth century, with a memorandum to decision-makers.* Paris: UNESCO.

Boli, J., and F. O. Ramirez. 1992. Compulsory schooling in the Western cultural context. In *Emergent issues in education: Comparative perspectives,* edited by R. F. Arnove, P. G. Altbach, and G. P. Kelly. Albany, NY: State University of New York.

Bourdieu, P., and L. Wacquant. 1992. *An invitation to reflexive sociology.* Chicago: University of Chicago Press.

Brandão, C. R. 1980. *A questão política da educação popular.* São Paulo: Brasiliense.

Cairns, J. 1975. MOBRAL—the Brazilian literacy movement: A first-hand appraisal. *Convergence* 8 (2), 12–23.

Coraggio, J. L. 1996. *Desenvolvimento humano e educação.* São Paulo: Cortez Editora.

Costa, M. A., V. Jaccoud, and B. Costa. 1986. *MEB: Uma história de muitos.* Rio de Janeiro: Vozes/NOVA.

Costa, M. V., ed. 1998. *Educação popular hoje.* São Paulo: Edições Loyola.

di Ricco, G. M. J. 1979. *Educação de adultos: Uma contribuição para seu estudo no Brasil.* São Paulo: Edições Loyola.

Easton, P., and S. Klees. 1992. Conceptualizing the role of education in the economy. In *Emergent issues in education: Comparative perspectives,* edited by R. F. Arnove, P. G. Altbach, and G. P. Kelly. Albany, NY: State University of New York Press.

Eshiwani, G. 1989. The World Bank document revisited. *Comparative Education Review* 33 (1): 116–129.

Fernandes, C., and A. Terra. 1994. *40 horas de esperança: O método Paulo Freire. Política e pedagogia na experiência de Angicos.* São Paulo: Editora Ática.

Fink, M., and R. F. Arnove. 1991. Issues and tensions in popular education in Latin America. *International Journal of Educational Development* 11 (3): 221–230.

Fonseca, M. 1996. O financiamento do Banco Mundial á educação Brasileira: Vinte anos de cooperação internacional. In *O Banco Mundial e as políticas educacionais,* edited by L. d. Tommasi, M. J. Warde and S. Haddad. São Paulo: Cortez Editora.

Freire, P. 1970. *Pedagogy of the oppressed.* New York: Continuum.

Gadotti, M. 1994. *Reading Paulo Freire: His life and work.* Albany: State University of New York Press.

Gauri, V. 1998. *School choice in Chile: Two decades of educational reform.* Pittsburgh: University of Pittsburgh Press.

Ghiraldelli, P. 1994. *História da educação* (2nd ed.). São Paulo: Cortez.

Graciani, M. S. 1997. *Pedagogia social de rua.* São Paulo: Cortez.

Gramsci, A. 1971. *Selections from the Prison Notebooks,* edited and translated by Q. Hoare and G.N. Smith. New York: International.

Harbison, R. W., and E. A. Hanushek. 1992. *Educational performance of the poor: Lessons from rural Northeast Brazil.* New York: Oxford University Press, for the World Bank.

Hausmann, C. 1998. *Nonformal education for women in Zimbabwe : Empowerment strategies and status improvement.* Frankfurt: Peter Lang.

Henales, L., and B. Edwards. 2000. Neoliberalism and educational reform in Latin America. *Current Issues in Comparative Education* 2 (2). Available at: http://www.tc.columbia.edu/cice/articles/lhbe122.htm

Holland, D., and J. Lave. eds. 2001. *History in person: Enduring struggles, contentious practice, intimate identities.* Santa Fe, NM: School of American Research Press.

Holland, D., C. Lutz, D. Nonini, L. Bartlett, T. Guldbrandsen, M. Fredericks, and E. Murillo, Jr. 2002. *If this is democracy.* Manuscript submitted for review and publication.

Hollanda, E. 1997. O Mobral de Dona Ruth. *Istoé.* May 28. Available at: http://www.terra.com.br/istoe/politica/144316.htm. [Accessed 30 August 2002.]

Jannuzzi, G. S. d. M. 1979. *Confronto pedagógico, Paulo Freire e Mobral.* São Paulo: Cortez and Moraes.

Jones, P. 1990. UNESCO and the politics of global literacy. *Comparative Education Review* 34 (1): 41–60.

Jones, P. W. 1992. *World Bank financing of education: Lending, learning, and development.* New York: Routledge.

Kane, L. 2001. *Popular education and social change in Latin America.* London: Latin American Bureau.

Kindervatter, S. 1979. *Nonformal education as an empowering process with case studies from Indonesia and Thailand.* Amherst: Center for International Education, University of Massachusetts.

Labaree, D. 1997. *How to succeed in school without really learning: The credentials race in American education.* New Haven: Yale University Press.

La Belle, T. J. 1986. *Nonformal education in Latin America and the Caribbean.* New York: Praeger.

Lemos, F. d. A. 1996. *Nordeste: O vietnã que não houve. Ligas camponesas e o Golpe de 64.* Londrina, Paraná: Universidade Estadual de Londrina.

Levy, D. C. 1986. *Higher education and the state in Latin America: Private challenges to public dominance.* Chicago: University of Chicago Press.

Lovisolo, H. 1990. *Educação popular: Maioridade e conciliação.* Salvador: UFBA/Empresa Gráfica da Bahia.

Meyer, J. W., and F. O. Ramirez. 2000. The world institutionalization of education. In *Discourse formation in comparative education,* edited by Jürgen Schriewer. New York: Peter Lang.

Meyer, J. W., J. Boli, G. M. Thomas, and F. O. Ramirez. 1997. World-society and the nation-state. *American Journal of Sociology* 103 (1): 144–181.

Nunes, M. de L. 1992. *Alternative education and social change in Brazil: A history and case study.* Ph.D. dissertation, Louisiana State University and Agricultural and Mechanical College.

Omi, M., and H. Winant. 1994. *Racial formation in the United States: From the 1960s to the 1990s.* New York: Routledge.

Oxenham, J. 1975. *Nonformal education approaches to teaching literacy.* East Lansing: Institute for International Studies in Education, Michigan State University.

Paiva, V. P. 1973. *Educação popular e educação de adultos.* São Paulo: Edições Loyola.

Paiva, V. P. 1980. *Paulo Freire e o nacionalismo-desenvolvimentista.* Rio de Janeiro: Civilização Brasileira: Edições UFC.

Plank, D. 1987. The expansion of education: A Brazilian case study. *Comparative Education Review* 31 (3): 361–375.

Plank, D. N. 1996. *The means of our salvation: Public education in Brazil 1930–1995.* Boulder, CO: Westview Press.

Puiggros, A. 1996. World Bank education policy: Making liberalism meet ideological conservatism. *NACLA Report on the Americas* 29 (6): 26.

Puiggros, A. 1999. The consequences of neoliberalism on the educational prospects of Latin American youth. *Current Issues in Comparative Education* 1 (2). Available at: http://www.tc.columbia.edu/cice/articles/ap112.htm

Scocuglia, A. C. 1997. *A história da alfabetização política na paraíba dos anos sessenta.* Recife: Universidade Federal de Pernambuco.

Sheffield, J., and V. P. Diejomaoh. 1972. *Nonformal education in African development.* New York: African-American Institute.

Singh, R. Pal. 1987. *Nonformal education: An alternative approach.* New Delhi: Sterling Publishers.

Souza, J. F. de, and Z. Granja Porto, eds. 2000. *Educação popular: Participação, exclusão na América Latina hoje.* Recife: NUPEP.

Srinivasan, L. 1977. *Perspectives on nonformal adult learning.* New York: World Education.

Steiner-Khamsi, G. 2000. Transferring education, displacing reforms. In *Discourse formation in comparative education,* edited by Jürgen Schriewer. New York: Peter Lang.

Stromquist, N. P. 1997. *Literacy for citizenship: Gender and grassroots dynamics in Brazil.* Albany: State University of New York Press.

Sutton, M., and B. Levinson, eds. 2001. *Policy as practice: Toward a comparative sociocultural analysis of educational policy.* Stamford, CT: Ablex Publishing.

Torres, C. A. 1995. *Education and social change in Latin America.* Albert Park, Australia: James Nicholas Publishers.

UNESCO. 1975. *MOBRAL—The Brazilian Adult Literacy Experiment.* UNESCO Regional Office for Education in Latin America and the Caribbean.

UNESCO. 1986. *Formal and nonformal education: Co-ordination and complementarity.* Bangkok: UNESCO Regional Office for Education in Asia and the Pacific.

Williams, R. 1977. *Marxism and literature.* Oxford: Oxford University Press.

Winant, H. 1994. *Racial conditions: Politics, theory, comparisons.* Minneapolis: University of Minnesota Press.

Yadav, S. K. 1987. *Nonformal education: New policy perspective.* New Delhi: Shree Publishing House.

CHAPTER NINE

EUROPEANIZATION AND FRENCH PRIMARY EDUCATION

Local Implications of Supranational Policies

Deborah Reed-Danahay

There is now a world culture, but we had better make sure we understand what this means: not a replication of uniformity but an organization of diversity.

(Hannerz 1996:102)

MASS EDUCATION AND THE "NEW EUROPE"

The European Union takes a growing role in social policy, including educational policy, among its fifteen member nations. The creation of a "New Europe" requires, according to most of its architects, the creation of a European identity and notions of European citizenship. We are, consequently, seeing new educational policies and reforms incorporating what is called a "European dimension" in schools. A new sort of European citizen or "self" is the desired outcome of such educational changes: one who identifies with a wider social space beyond the nation. As Maryon McDonald has noted, beginning in the mid-1980s EU officials came to realize that the economic models of the EU were not sufficient for the creation of a "People's Europe"; these officials came to believe that "people needed to know they were European . . . and such elements as the European flag and an anthem were introduced" (McDonald 1996:54). The role of education in the construction of Europe is expanding rapidly, particularly since the explicit mention of education in the Amsterdam Treaty of 1997. The treaty charges in Article 149 that Community action shall be aimed at

> [d]eveloping the European dimension in education, particularly through the teaching and dissemination of the languages of the Member States; encouraging mobility of students and teachers, inter alia by encouraging the academic recognition of diplomas and periods of study; promoting cooperation between educational establishments; developing exchanges of information and experience on issues common to the education systems of the Member States; encouraging the development of youth exchanges and exchanges of socio-educational instructors; encouraging the development of distance education.

This trend raises important questions about the relationship between local, national, and supranational influences upon education. It also raises questions about the lived experiences of pupils, teachers, and families as new structures and discourses of schooling emerge.

The "mass education and world culture" approach of John Meyer, John Boli, Francisco O. Ramirez, and others argues that a world culture model of education arose in Europe beginning with the Enlightenment and has since spread throughout the globe (Boli and Ramirez 1986; Boli and Ramirez 1992; Boli, Ramirez, and Meyer 1985; Ramirez and Ventresca 1992). Compulsory mass education, according to these authors, is connected to the rise of the nation-state; and its role has been to create "modern" citizens in these geopolitical units. "Transnational masses" were, thus, transformed through schooling into national citizens (Ramirez and Ventresca 1992:50). It is, therefore, timely to "return" to Europe and to contemporary reforms in European education in light of this model. In contemporary Europe, "national citizens" are encouraged to transform themselves into "European citizens."

It is particularly relevant to address processes of Europeanization and schooling in France, a nation famous for its use of schools to produce French citizens in the nineteenth and early twentieth centuries. In this essay, I will focus on two projects funded by the European Union in one French department—the Puy-de-Dôme, located in the Auvergne region of central France.[1] These projects, "European Folktales" and "New Horizons," had different funding structures and different aims, yet both shared an emphasis on computer technology and on exchanges between pupils in different European regions and nations. These projects reflect reforms in French education in which the supranational influence of Europe is paramount. A focus on national identity, so crucial to early aims of the French primary school system, is notably absent from the content of these projects: One emphasizes European identity; the other emphasizes regional identity. In one of these projects, English was the sole language of communication among pupils in different nations (none of which was an Anglophone country!). There is a great irony in identifying such a project with France, a country famous for its imposition of a national language over regional languages in order to create "the nation."

These projects lead me to question the usefulness of the mass education and world culture model in understandings of contemporary developments in European education. For the proponents of this model, the school as an institution is taken for granted, in that its existence seems a "natural" development in the rise of the nation-state and it assumes a somewhat "neutral" role in enhancing the aims of the nation-state—primarily that of gaining legitimacy on the world stage, presumably in order to attain or sustain economic and military strength. I will eventually outline below another approach, that of Michel Foucault, which problematizes the institution of education in what I see as a more fruitful way of understanding contemporary schooling. Before doing so, however, I would like to call attention to two main assumptions upon which the mass education and world culture model rests but that do not bear out in the context of contemporary Europe. These are the centrality of the nation-state and the lack of coercion in the adoption of converging educational reforms.

According to the mass education and world culture model, there are three major components of mass education: (1) an emphasis on universal and standardized compulsory education; (2) a high level of institutionalization (including national ministries of education); and (3) an emphasis on the socialized child-learner as a "modern" individual. The unit of the nation-state is crucial to this model, because it was in the context of nation-state formation that mass education arose. In their earlier competition within the "interstate system" of Europe, individual European nations adopted state-sponsored mass education in order to legitimize their political and economic positions (Boli and Ramirez 1986; Ramirez and Boli 1987). Emerging nation-states outside of Europe and North America, the argument goes, accept this institutional form because they see the advantages for establishing their legitimacy in "looking modern" (Fuller 1991; see also the critique of Anderson-Levitt and Alimasi 2001). A key element for Meyer and his colleagues is the lack of coercion in the spread of this educational form: Mass education is adopted because it fits with certain shared ideals about citizenship, progress, and the individual.

Even though all European nations have adopted national systems of public primary education, as Meyer and colleagues argue, the mass education and world culture model is undermined by the presence of a great deal of diversity among these systems in terms of philosophy of education and methods of education. This is a pivotal issue in the EU right now. Because of the strong association between primary education and nation building in Europe, most EU countries resist formal directives about primary education coming from the EU and want to maintain autonomy in this realm. Each nation sees its own primary education system as uniquely reflecting national values. At the level of primary education in Europe, there are no "European" standards of curriculum or mandatory educational policies of any sort. If one looks at the official programs and rhetoric associated with the "European dimension of education," change is not expected to occur through explicit directives or the imposition of uniform standards across different systems with different histories. This is particularly true at the level of primary education. Thus, "Eurocrats" propose that convergence among the various educational systems will be achieved through a "free market" exchange of ideas—including school partnerships, teacher exchanges, comparisons of educational systems (see, for example, Aubray and Tilliette 2000). The best of each system will somehow float to the top and be adopted rationally by each other system—very much a neoliberal view. This is in keeping with the Meyer group position that the spread of mass education occurred not primarily through coercion; rather, it came to be seen as something of value.

But change in European education is not occurring via a "natural" process of like-minded people seeing things in the same light, despite what both the Meyer group and the Eurocrats rosily argue. Rather, certain initiatives and ideas are being fostered through grants to schools and other subsidies. Resources are being allocated to those who conform to the new ideals, which is a form of coercion. This will become apparent shortly as I turn to the ethnographic examples I will present below. European educational subsidies are in keeping with a wider global trend toward the increasing role of subsidy in policy reform and implementation, operating through a process in which "government does not itself carry out the activities but instead seeks to achieve its goals by influencing the behavior of subsidy-recipients" (Leeuw 1998:79).

European educational initiatives also call into question the necessary association between the nation-state and mass education. Although Boli and Ramirez (1986) draw attention to the interstate system, in which ideas about education and nationhood emerged, showing the wider circulation of ideas and strategies, the increasing influence of *Europe*—the EU—undermines national autonomy in crucial ways. Recent reforms and initiatives in education, common economic policies, and a common currency are among the factors contributing to the weakening of the nation-state in Europe, with *regions* growing in importance. Despite the resistance among nationalists in Europe to EU initiatives (as in the case of educational reform noted above), the influence of the EU in all features of life in Europe is now deeply rooted.

In order to understand the role of education in contemporary Europe, and the meaning of schooling reforms at the local level, it is necessary to unpack the "controlling processes" (Nader 1997) of education. The approach of Michel Foucault, which emphasizes discourse and techniques or disciplines of power and control, is useful in this enterprise. The discourse of "education" as a modern and therefore desirable thing, part of the Meyer group's argument, is seen in a different light from the perspective of Foucault. The projects that I will describe below operate within a discourse in which things European and the use of technology are symbols of being "modern." As a middle school teacher in Clermont-Ferrand, explaining to me the attraction of incorporating the European dimension into French schools, put it: "Europe symbolizes modernity."

Foucault (1977) has contributed insightful analyses of the way in which modern Western societies regulate thoughts and behavior. For Foucault, power works in modern society (post-eighteenth century) through "disciplines"—techniques and knowledge that regulate through surveillance of individuals. Those in positions of domination work through tactics and strategies to "discipline" people to behave and think in certain ways. Foucault's is not a top-down approach to power, however, but one that indicates how individuals come to regulate and monitor their own behaviors as they internalize discipline. This works through the collection of information about individuals (for example, in school files), surveillance of individual behavior (checkpoints, supervisors), and control over bodies (standing in straight lines, sitting still in class). Foucault focused on key institutions such as the prisons, hospitals, and schools. As Stephen Ball (1990) writes, Foucault's focus was on discourses that control meaning and constrain ways of thinking. "Education institutions control the access of individuals to various kinds of discourse" (Ball 1990:3). Foucault's theory enlarges the view of mass education to show that schools work in concert with other institutions to produce a particular sort of modern "subject." Foucault is also useful for thinking about education because of his emphasis on the inextricable relationship between power and resistance.

As Shore and Wright (1997) point out in an essay on anthropological approaches to policy, a key task of governments in neoliberal democracies is to use "techniques of the self" (a Foucauldian concept) to create a self-regulating subject. The EU projects described here ultimately work to create new forms of "moral regulation" through their use of subsidies for technology and exchanges between students in different regions or nations within Europe. It is not so much the overt content of each project that works toward identity construction, but, rather, the social relations and social practices encouraged.

EUROPEANIZATION AND FRENCH EDUCATION

The two cases that I will now describe were funded by subsidies from the European Union through grant competitions. The two projects both involved partner schools in other EU nations, but they originated in a relatively marginal region in France, the department of Puy-de-Dôme, in Auvergne. I have chosen to focus on this region so distant from the Parisian center of France in order to illustrate the ways in which social actors in the hinterlands may take advantage of policies generated far away for their own purposes at the same time that they (inadvertently) come under new power arrangements as they do so. In this and my previous work (Reed-Danahay 1996) in this region, I follow what Michael Herzfeld has recently referred to as "the characteristic stance" of anthropology in "its proclivity for taking marginal communities and using their marginality to ask questions about the centers of power" (Herzfeld 2001:5). The European Union is, itself, interested in such marginal regions, and devotes a great deal of time and energy to formulating programs aimed at "development" in hinterland regions of Europe. While France may not strike the reader at first glance as a place to examine marginality in Europe, being so central historically to the formation of the European Union, several regions in France (including portions of Auvergne) are targeted by the EU for special subsidies due to their underdevelopment. Approximately 20 percent of all townships (*communes*) in France are rural. In the Auvergne region, which is one of the most rural areas of France, twice as many people (8 percent vs. 4 percent) are full-time farmers than in the rest of France. About half of all the townships in the Puy-de-Dôme are rural townships, eligible for subsidies from the EU. It is into these more remote regions of Europe that the EU has to work hardest to penetrate; urban areas are already more easily under the gaze of emerging institutions of the EU.

Case 1A: Comenius-funded Project

This project was in its third year when I met with its director in the spring of 2001. This project involved three schools in France, one in Italy, one in Norway, and one in Germany. The main theme was a comparison of folktales common to these European countries, with a major final project on the tale of Sleeping Beauty. Funding for this initiative came from Comenius,[2] a European Commission program for school projects. The European Commission is the institution within the European Union that proposes policies and legislation and ensures implementation of Treaty provisions.

Before getting to the specifics of this project, I will briefly outline the organizational structure under which the Comenius program is administered in France, starting at the level of the individual school. Each EU nation has a different structure for administering this program. French public elementary schools (and private schools under state contract) are under the jurisdiction of local and regional authorities. At its most basic level, the French administrative structure is organized by geographical units of increasing size—*commune* (i.e., municipality), department, region, and nation. *Communes* provide the buildings and most materials used in the classrooms, although the national educational system controls the curriculum and employs the teachers, who are civil servants. Direct national control over primary schools is through the primary education inspectors, each of whom has a particular section of a department to supervise. Inspectors are, in turn, supervised by the rector of the regional Académie, the unit responsible for primary through high school education in France. Universities are under direct national control. Inspectors ensure that schools conform to national standards of curriculum and also serve as advisers to teachers.

In each Académie, there is an office of international programs (Délégués Académiques aux Relations Internationales et à la Coopération, DARIC) charged with coordinating international initiatives in education, including projects of the European Union. The director of this office receives announcements of funding possibilities, and is supposed to work with the inspectors to advertise the news among school directors. During my fieldwork, I came to know the school inspector responsible for this project (who was also the school inspector for the schools in the other project I will discuss below), and interviewed the DARIC staff person responsible for the Socrates program, which is the umbrella program under which Comenius is housed.

Each national government in the EU provides an agency to accept and review proposals that are then funded by the EU. There is much discretion permitted as to how the nation decides to handle this. In France, the Comenius program and all other education programs funded by the European Commission are administered through the Socrates Agency, located in Bordeaux. During my fieldwork, I visited this agency and interviewed staff. The employees of this agency are not employed directly by Europe; they are either French civil servants or contract personnel. All French proposals for the

Comenius program pass through this agency, and also must be approved at EU headquarters in Brussels. There are relatively few projects at the elementary level in France funded by the agency; most are at the secondary level. The staff at the agency told me, however, that most of the elementary school projects are located outside of major cities. They explained that in large urban areas, schools can get subsidies from city government and do not need to seek funding sources elsewhere (such as Europe). Therefore, it is fairly typical to see a project originating in France in a small town or village.

The director of the European Folktales Project is the principal of a primary school complex in a working-class town on the outskirts of Clermont-Ferrand. This school has impressive computer resources compared with other primary schools I visited in this region. About 20 shiny new Macintosh computers were arranged in a computer room, their boxes still in evidence. Dynamism in this school was also apparent in the visual displays of creative student work throughout the halls. The Folktales Project had been organized through planning meetings among personnel in each school that had taken place at various locations (one in Norway, for instance). The children did not travel, but exchanged information about themselves with each other through email and through postal mailings of their work. Children in each school filled out questionnaires about their everyday lives, for example, and drew pictures of their towns or villages to share with other schools. Partnerships among the six schools developed. The language of communication was English in all cases. At the end of the project, each participating class (in some cases, more than one class at each school participated) produced a version of the tale of Sleeping Beauty. Most of these were illustrated books (in English) but one class did a performance on video.

When I asked the director of this project what he felt were its aims, he spoke of the need to develop concepts of European citizenship among children, of the need for educational programs that led to an opening of their minds (*l'ouverture de l'esprit*). He felt that this was important for the creation of a new Europe free of racism and xenophobia, a Europe of peace. If European children could see that they shared a common cultural base, this would be a step toward that direction. I also asked the director to explain how he came to initiate this project and learn of the EU funding. He told me that he had seen a notice about funding in the *Bulletin Officiel*, the French official publication for education that informs teachers and administrators of programs and initiatives, curriculum, schedules, and the like. He had already been experienced in grant writing in his work on computer technology. And in his discussions with a former school inspector, he had already begun to think about using Greek myths to teach children about European heritage.

When he saw the notice about the Comenius program, he began to think about ways to develop this idea and draw upon previous networks to find school partners in other EU countries. At a conference that he had attended in Paris on education and European citizenship, he had made some connections, and he contacted an Italian school that he had previously been in touch with on another project.

The director of this project is very entrepreneurial and open to pursuing opportunities such as the Comenius funding. This case shows the necessity of an entrepreneurial spirit in obtaining resources in the current EU climate. It also demonstrates the director's abilities to gather support from many levels. As a media specialist and principal, the director has more time than the average teacher to devote to this, since he is not always in the classroom. He received the support of national education bureaucrats (*décideurs*, or decision makers, as they are now more euphemistically termed in France). Both his former inspector and the current inspector share his vision of a New Europe and are not opposed to the idea of using English as the primary language in such projects. Support also came from the administration of the *commune* in which the school was located, which shares a European outlook. The town has signposts in it that indicate distance from various European cities, and it considers itself a "*commune* of Europe." A sign proclaiming this designation is at the entrance to the town. As I interviewed the director one afternoon, a group of children playing baseball with the PE teacher were in view through the windows looking out on the large school grounds. Baseball is not, as one might imagine, a game commonly played in rural French schoolyards.

The strength of support for this project from other teachers and parents is less clear. Other teachers at this school were not heavily involved in this project, and teased the director during a coffee break when I was there that only he got to travel on the planning trips to other countries. I was not able to determine the exact number of students who participated in this project at the school (at least two classes participated), or the amount of time devoted to it in the classroom, since I arrived to study it toward the end of the school year. Although existing on a very small level, this project reflects a "unity in diversity" approach to education. Each school did its own project around a common theme, reflecting its own context. When I asked the teacher about the use of English, he did not feel that this threatened either French culture or language. For him, it was a practical matter, and this meta-level of communication could coexist with national and local culture.

This emphasis on common folktales in Europe, while creating bonds between children in different nations who share this heritage, also creates boundaries and raises troubling issues about who is and is not European. For instance, what would be the shared basis for commonality with immigrant children from Asia or North Africa in this scheme? Although I did not observe any such immigrant children at this particular school, there was a category of children there who were excluded by this project. While I was interviewing the director one day, a woman entered the room and was introduced to me. She worked with gypsy children at the school, who had separate classes due to language and other cultural difficulties as perceived by the educational authorities, and explained that there was an entire network of services provided to these children by the school district. These children were marginal at the school and did not participate in the European Folktales Project. I was quite struck by the disjuncture between the fact that the children in the mainstream were participating in a project aimed at creating a sense of shared identity with mainstream children in Germany or Norway while gypsy children in their own school were seen as "other." Although I did not directly pose this issue with the director, neither did he voice any recognition of this irony himself.

Case 2A: LEADER-funded Project

The LEADER project differed in several aspects from the Comenius project. First, it did not originate in the context of the school system per se, but from social actors involved in regional development projects. Second, it focused on regional rather than European identity. However, in many other ways, the projects display similar processes at work. The LEADER project, which was called "New Horizons," included schools in Italy, Normandy (a region in the north of France that borders the English Channel), and a deeply rural area of the Auvergne region of central France. This project was part of what is called the "regional policy" of the European Union, and not of educational policy in a strict sense. The schools received funding from the European Union through a project called LEADER, which is aimed at economic development at the regional level. According to my informants, this was the only project that has been funded by LEADER that involves schools. The project originated in Auvergne and was developed by a grouping of small municipalities (a *communauté de communes*), which employs staff (called "technicians") to develop tourism in the region and support the local economy through economic development projects. This organization had already organized previous projects in schools in the region with environmental themes. This time, it worked in partnership with an NGO that covered a wider geographical area and was also working on economic development projects funded by LEADER. As is mandated by most European funding sources, there were several partners in this project, as with the European

Folktales Project. The EU funding depended on there being more than one country involved, so the organizers found partners through an online database that exists for participants in LEADER initiatives.

As with any other locally initiated projects in France funded by Europe, this project had to pass through French administrative groups that vet projects for European programs such as LEADER. In this case, the unit was the administrative region of Auvergne, which approves LEADER projects and then passes these along to Brussels for final approval and funding. Because New Horizons was also involved in education, the school system had to be involved at a supervisory level. The school inspector had to be consulted about the project and of necessity became involved, even though the initiative for the project was completely local.

Like the Comenius project, the LEADER project involved school partnerships and exchange, a common theme for all participating schools, and use of the Internet. Here, the theme was the environment. The grant proposal included funds to equip each participating school with computer technology in order to achieve this part of the project. When I asked the technician who was a key actor in the conception of this idea to explain how it developed, she first explained the four main goals of the group of *communes* for which she worked. These were to support and maintain the agricultural base of the region, to manage forests, to manage urban (town) development, and to sensitize the local population to their own environment and its resources. The idea came to them to try to work through children in schools to enhance awareness of the environment among both parents and children. New Horizons was not the first project about the environment in schools that this grouping of *communes* had initiated; an earlier one had focused on water resources and another on pollution. It was, however, the first involving schools in other regions.

New Horizons was to involve children in understandings of their own region and its resources, who would then have exchanges with other children in other regions of Europe where students would also be learning about their own environment and sharing this knowledge with others. The project was, thus, conceived not so much as between nations but between regions in the European Union. Local schools were invited to participate and then partners were found (schools in Normandy and schools in the Frascati region of Italy). There was no particular design to the choice of partner regions, as a call had gone out to find partners and these were simply the ones that responded. The enticement to local teachers in Auvergne to participate was the offering of computers to their schools, all of which were small rural schools with meager resources.

The teachers and organizers met to plan this a few times, and several school classes were able to travel to the other regions. Visits and exchanges were made to Normandy and to Auvergne. Children from Normandy visited Auvergne and children from both Auvergne and Italy visited Normandy. No French children visited Italy, since there was resistance from parents in Auvergne for travel so far away. There were email exchanges between twinned schools for two years before the trips, which took place in the spring of 2001.

In addition to interviewing the technicians who organized this project and administered it, I visited two schools that had participated in the project. Children at both schools had traveled to Normandy with the project and come into contact with children from Italy while there. These schools presented a stark contrast to the school at which the Folktales Project was based. Both of these schools were lacking in many basic resources, particularly the smaller rural school. It was located in a tiny farm community, which one reached through twisting country roads. Almost all of the pupils were the children of farmers. This one-room school (referred to in France as a *classe unique*), in which children aged 3 to 11 were taught by one teacher and one aide, was not as much of an anomaly as it might appear. In the Puy-de-Dôme, 18 percent of all primary schools are one-room schools such as this. This school was located in a 100-year-old building, and the material conditions of the school were poor. The playground consisted only of a paved space to play, and there were no toys other than old tires and balls. The toilets were located in a separate building and were of the old-fashioned kind with merely a hole in the ground and periodic flushing of the water. Classroom furniture and materials were at a minimum. In one classroom, there were desks for the older children, but an old mattress on the floor was the place where the teacher gathered pupils for discussions, as when I came to visit.

The other school I visited was located in a small town and was swimming in a large building that used to house a middle school that had recently been closed. Twenty years ago, this town had a thriving Catholic middle school and a public middle school. A combination of depopulation and school restructuring had led to the disappearance of these two schools during the last decade. There were three classes in this school: one for younger children and two each combining two years of primary school. The class I visited (a combination of fourth and fifth grades) was taught by a teacher who had settled recently in this region but who was originally from the Paris area. The children in this classroom were a mixture of farm children, children whose parents worked in a neighboring spa town, and children of shopkeepers and other artisans.

These students had traveled to Normandy and had encountered Norman and Italian children. When I met them, they were drawing pictures describing the differences between the regions of Normandy and Auvergne and mounting photos from the trip for a display at an upcoming parent night. Many children expressed their view to me that the Norman children were snobby and the Italian children were very tied to home, since the Italian children all seemed to have cell phones and were frequently talking on the phone to people back home. The schools partnered with the Auvergne schools were in more affluent and urbanized regions. The Auvergne region is considered backward and isolated by many French, so it is not surprising that the Norman children would have stereotypical views of Auvergne. The children in Auvergne were, consequently, made to feel inferior through this exchange. The overall aim of the project was to help them value their region and its resources. An inadvertent consequence was that they saw new horizons and came to feel that "all there was" in Auvergne was nature: mountains, lakes, and forests.

The policy of the New Horizons project was to encourage exchanges between children so that they would come to value their own region and also become more open to other regions of Europe. The program faced several obstacles, besides that of the children's perceptions, outlined above. First, it depended upon email exchanges. Computers didn't arrive when expected, and several teachers felt they did not get enough training on how to use the computers. Because some teachers had participated in this project only because they wanted to get computers for their schools, they viewed this negatively. The educational authorities were supposed to offer training, and there was some disagreement about whether or not it was offered or offered in ways convenient for the teachers. Teachers, even those supportive of the project, felt that the project was a big burden on their time. Some teachers expressed the opinion that it was being imposed on them, but they took advantage of it for the sake of their students and the computer equipment they would receive. Translation was another problem. In contrast to the Folktales Project, conducted through the use of English, each country in the New Horizons project used its native language. French teachers, therefore, had to translate Italian and vice versa. The teacher in the one-room school was the daughter of an Italian immigrant and so she was able to easily translate for her students. Other teachers, however, complained about the communication problems presented in this type of exchange and told me that the school district had promised translation help that was never provided.

New Horizons was a local initiative but was not initiated by local teachers or parents. The role of Europe in this project was just as important as in the European Folktales Project, but at a different level. Here, the emphasis was not on producing a common sense of European citizenship through study of common myths and tales, but on the unique qualities of regions within the EU. The children's experience in the New Horizons project emphasized difference to a much greater degree than unity. In both cases, the nation of France and concepts of French national identity were oddly absent.

CITIZENS OF EUROPE, CITIZENS OF THE PLANET

In the case of the Comenius-funded project on European folktales, the students were being encouraged to see and share a common heritage with children in other European nations, and to consequently see themselves as Europeans. In the case of the LEADER-funded New Horizons project, the students were encouraged to value their own region and its resources and to see themselves as citizens of planet earth, in which other people in other regions preserved and valued their resources as well. The emphasis was not so much on a shared European identity; rather, it was on the uniqueness of each region within a shared framework of global ecological concern. As we can see, therefore, the aim of Europe to create citizens who identify themselves as European is complicated by the specifics of each project. While oppositions between national versus European identity were not posed by these projects, other types of opposition were reinforced through them. The European Folktales Project, as I suggested above, fostered a view of unity in diversity for European identity that depended upon a narrow idea of shared European heritage. Many citizens of Europe (immigrants and gypsies, for example) were excluded in this formulation. With the "New Horizons" project, identification with region and its resources took precedence over notions of shared identity. Regional difference and distinctiveness was in the foreground, and the unity was of the natural resources of the planet, not of Europe as a cultural category. These cases call into question the claim by Boli and Ramirez (1992) that "transnational masses" are transformed into national citizens through schooling. This is a macro-level approach that ignores the other perspective "on the ground," in which schools attempt to turn people with very local identities—highly differentiated according to local languages, values, economic systems, and so on—into a "mass" with a shared national identity. With these two projects, we see schools attempting to reinforce regional identity on one hand, and de-emphasizing the local or national in favor of the transnational or global on the other.

Both of these programs adopted school partnerships among schools in France and other European nations and student exchange of information through computer and Internet technology as a pedagogical tool. Although the two programs operated under different institutional structures, utilized different subject matter, and were located in different sociocultural contexts, they were both aimed at producing a new type of "European" pupil and citizen. If we look at the techniques to achieve this, rather than at the content or even administrative structure of each project, we can see similarities. The association of technology with modernity and with a "modern" Europe, as well as the use of exchanges between pupils and schools, is part of a discourse about harmonization among different nations and national systems of education. Supranational education, like mass education, is being spread to "the masses" in ways that appear to lack coercion. Students are being encouraged to "think European" (Aubray and Tilliette, 2000:35, cited above), rather than to stop being French or Auvergnat.

These cases show that the issue of increased standardization and globalization in education proposed by the mass education and world culture theory is, perhaps, phrased in a way that does not adequately capture what is happening in Europe today. Looking primarily at quantitative measures such as length of school year and of school day and hours spent in various aspects of the curriculum misses more subtle aspects of reform. Rather, a focus on a convergence of techniques for achieving order and discipline, a more Foucauldian approach, is also needed in order to address what one can see in the two cases described above and no doubt other examples of education at the local level. New tactics of surveillance, new subjectivities, and new disciplinary techniques are at work in the adoption of a European dimension of education. Whether this is initiated through the national ministry of education, the European Union, local teachers, or NGOs, there is a convergence of ideas about the need for mobility and exchange, and for the use of new technologies in education in France today. People are thus instructed "to be" in new ways in the world. Children and their teachers in isolated rural schools or schools in small towns are thereby encouraged to interact with wider frames of reference, facilitated by new modes of communication via the Internet and improved modes of travel. This interaction goes both ways, however. Schools wanting subsidies for technology needs in their schools become caught in the web of record keeping and surveillance by the agencies that fund their projects. This penetration of supranational levels of surveillance into schools in the most remote sections of rural France suggests a new form of control beyond that of the already centralized French state. While the content of the projects sends various messages to children about identity and confers knowledge of this content (whether it be scientific, cultural, literary, or other), it is the form of the projects that requires examination.

The mass education and world culture approach of Meyer, Boli, Ramirez, and others is an important framework for thinking about phenomena associated with the spread of national or transnational systems of education and convergences in their policies, values, and curricula. The world culture approach helps our understandings of the rhetoric of European policymakers and the ways in which convergence may happen without mandatory reforms. Foucault's emphasis on discipline and governmentality permits a different level of understanding of such processes, however—one that does not take the institution of education as a "natural" social form but as one that is embedded in discourses of power and control. We must, therefore, look at the ways in which what goes on in schools is part of a "network of technologies of surveillance and observation" (Rousmaniere, Delhi, and Coninck-Smith 1997) connected to forms of moral regulation. The new moral discourse involves new technologies of pedagogy—groups of schools working on common projects using computers and forms of exchange. This involves cooperation within a wider universe of competition for the very subsidies that fund the projects and that work to ensure the implementation of these new technologies. We also see that it requires entrepreneurial social actors who will seek this funding, and that all participants (especially teachers themselves) are not always enthusiastic accomplices in these initiatives.

NOTES

1. This ethnographic research was funded by a Fulbright Research Award. I spent three months researching EU programs in schools (May-July 2001.) I returned to France for further research on the Europeanization of primary education under the generous sponsorship of the Institut National de Recherche Pédagogique in Paris during June and July 2002. I have conducted fieldwork in the Auvergne region of France for over twenty years.

2. Comenius is part of the Socrates program of the EU. These programs encourage and fund projects for children and teachers in member states and potential member states at primary and secondary levels of education. Comenius deals primarily with primary education, while other programs deal with secondary and higher education and teacher training. John Amos Comenius, for whom the project is named, was a Moravian (Czech) philosopher in the seventeenth century who argued for collaboration in universal education and even suggested the formation of an international ministry of education (see Cauly 2000; European Commission 2001).

REFERENCES

Anderson-Levitt, K. M., and N.-I. Alimasi 2001. Are pedagogical ideals embraced or imposed? The case of reading instruction in the Republic of Guinea. In *Policy as practice: Toward a comparative sociocultural analysis of educational policy*, edited by M. Sutton and B. A. Levinson. Westport, CT: Ablex.

Aubray, B., and B. Tilliette. 2000. La mondialisation éducative. *Cahiers d'Europe* 3: 30–35.

Ball, S. J. 1990. Introducing Monsieur Foucault. In *Foucault and education: Discipline and knowledge*, edited by S. J. Ball. London: Routledge.

Boli, J., and F. O. Ramirez. 1986. World culture and the institutional development of mass education. In *Handbook of theory and research for the sociology of education*, edited by J. G. Richardson. New York: Greenwood Press.

Boli, J., and F. O. Ramirez. 1992. Compulsory schooling in the Western cultural context. In *Emergent issues in education: Comparative perspectives*, edited by R. F. Arnove, P. G. Altbach, and G. P. Kelly. Albany: SUNY Press.

Boli, J., F. O. Ramirez, and J. W. Meyer. 1985. Explaining the origins and expansion of mass education. *Comparative Education Review* 29 (2): 145–170.

Cauly, O. 2000. Comenius: fondements pour une éducation nationale et universelle. *Cahiers d'Europe* 3: 36–54.

European Commission. 2001. Comenius: European cooperation on school education. http://europa.eu.int/comm/education/socrates/comenius

Foucault, M. 1977. *Discipline and punish: The birth of the prison*, translated by A. Sheridan. New York: Pantheon Books.

Fuller, B. 1991. *Growing up modern: The Western state builds Third-World schools.* New York: Routledge.

Hannerz, U. 1996. *Transnational connections.* London: Routledge.

Hertzfeld, M. 2001. *Anthropology: Theoretical practice in culture and society.* Malden, MA: Blackwell.

Leeuw, F. L. 1998. The carrot: Subsidies as a tool of government—theory and practice. In *Carrots, sticks and sermons: Policy instruments and their evaluation*, edited by M.-L. Bemelmains-Videc, R. C. Rist, and E. Vedung. New Brunswick, NJ: Transaction Publishers.

McDonald, M. 1996. "Unity in diversity": Some tensions in the construction of Europe. *Social Anthropology* 4 (1): 47–60.

Nader, L. 1997. Controlling processes: Tracing the dynamic components of power. *Current Anthropology* 38 (5): 711–737.

Ramirez, F. O., and J. Boli. 1987. The political construction of mass schooling: European origins and worldwide institutionalization. *Sociology of Education* 60 (1): 12–17.

Ramirez, F. O., and M. J. Ventresca. 1992. Building the institution of mass schooling: Isomorphism in the modern world. In *The political construction of education: The state, school expansion, and economic change*, edited by B. Fuller and R. Rubinson. Westport, CT: Praeger.

Reed-Danahay, D. 1996. *Education and identity in rural France: The politics of schooling.* Cambridge: Cambridge University Press.

Rousmaniere, K., K. Delhi, and N. de Coninck-Smith. 1997. Introduction. In *Discipline, moral regulation, and schooling: A social history,* edited by K. Rousmaniere, K. Delhi, and N. de Coninck-Smith. New York: Garland Press.

Shore, C., and S. Wright. 1997. Policy: A new field of anthropology. In *The anthropology of policy: Critical perspectives on governance and power,* edited by C. Shore and S. Wright. New York: Routledge.

CHAPTER TEN

TRANSFORMING THE CULTURE OF SCIENTIFIC EDUCATION IN ISRAEL

Kalanit Segal-Levit

This chapter deals with Israeli adoption and application of a culture of scientific education that previously existed in the former USSR. It presents the development of an educational initiative that originated among scientists and teachers who had previously taught in Soviet schools for excellence in mathematics and physics.

The initiative began with the establishment of a complementary educational structure that answered immigrants' need for an education that would preserve their culture in science and in other subjects (such as English, computer sciences, chess, and Russian). Certain conditions in Israeli society, in interaction with the characteristics of the wave of immigration from the former USSR to Israel in the 1990s, enabled the initiative to develop into a structure that brought together immigrants, veteran teachers, and students in the formal educational system. The penetration and expansion of this initiative into science classes within the formal educational system transformed and hybridized the original culture of scientific education. The processes of feedback and interactions that developed in the schools influenced the general educational atmosphere as well.

THE RESEARCH QUESTION

Immigrants usually find themselves to be a marginal group, placed on the fringes of society. Eisikovits (1995a) defines immigration as an acute situation of cultural change. Much research has dealt with the difficulties that characterize the integration of immigrants into various educational systems. Gibson (1995), Ogbu (1991), and Ogbu and Simons (1998) explain the success of immigrants at school in terms of the immigrants' perception of their chances of integrating into the job market. According to these authors, belonging to a certain minority group shapes the group's perception of its status and of its chances for integration with the establishment and the mainstream.

The current chapter deals with a unique phenomenon, a borrowing of the immigrants' culture of scientific education by the formal educational system. The chapter follows the transformation that took place in the teaching of mathematics, physics, and other subjects in schools that employed teachers who belong to the "Top" organization. ("Top" is a pseudonym.) As I will explain below, the Top organization first offered after-school classes in Russian, then offered its alternative program in Hebrew, and now has come to be integrated into the regular school system. Especially fascinating is the role of immigrants as the initiators of this educational change.

I will present in this chapter the factors that caused the immigrants to develop the new structure. I will also analyze the conditions in Israeli society that enabled its foundation and development into an Israeli–immigrant system that penetrated the formal educational system in the form of science classes. The analysis of the phenomenon will explore its place in relation to both globalization and multiculturalism, in Israeli society in general and within its educational system.

The chapter is based on a two-year-long ethnographic study that included observations in ten schools where the Top organization operated, focusing on two. The observations included lessons, recesses (both in the teachers' room and among the pupils), informal activities such as trips, special activities, staff meetings, parents' days, and individual meetings of the staff with pupils and their families. Observations were also held during the staff meetings of the organization, as well as during meetings of the organization's general manager with the school staffs, with people in the educational, municipal, and political establishment, and with potential funding sources. I also held interviews with parents and pupils, with principals, and with staff at the organization and at the schools where it operated. Interviews were also held with persons in the educational establishment, both municipal and political, and with persons who were connected with the organization's activities. I checked information gathered during interviews in correspondence with functionaries in the various government offices, the Knesset, the educational systems, and the municipal establishment.

THE EDUCATIONAL SYSTEM:
CONDITIONS FOR THE DEVELOPMENT
OF AN EDUCATIONAL INITIATIVE

Two aspirations of the educational establishment in Israel are relevant to our case. The first is the aspiration for equality that exists throughout the history of Israel's educational system. This aspiration embodies aspects of "institutional theory" as defined by Fuller (1991) or "global rationalization" as defined by Davies and Guppy (1997).

The starting point of this aspiration is the similarity between central and peripheral countries in the modern institutions that represent centralized authority. Whatever their level of social development, countries claim authority and responsibility over many aspects of modern life, including education (Fuller and Rubinson 1992; Meyer 1980; Thomas, Meyer, Ramirez, and Boli 1987). The educational policy that was formed in Israel, since its foundation, was expressed by the establishment of a state educational system representing the state's central authority and responsibility. This responsibility is also expressed by the laws providing for compulsory, free schooling that were meant to supply mass education. According to Ramirez and Ventresca (1992), mass education is a national project that assumes validity on the level of a global political culture that defines education as a basic human right for all citizens.

According to this approach, citizenship links individuals not only to the state as a bureaucratic organization, but more importantly, to an "imaginary community." Mass education has become a central array of activities through which reciprocal relations between individuals and the state are shaped. Aspiration to use the educational system for the socialization of future citizens stems from ideologies that shaped the organizational structure of the educational system (Benavot, Cha, Kamens, Meyer, and Wong 1991). Since Israel sees itself as a country of immigrants, the aspiration among Israelis was to form a uniform model of the *tzabar* (native-born Israeli) by assimilating the various cultural variants brought by immigrants of various origins, through a "melting pot" policy. According to Zionist ideology, the figure of the native pioneer who works the land, the *tzabar*, was the only legitimate cultural hero. In the interest of the effort to shape a new nation, it negated the legitimacy of other cultural styles (Eisikovits 1997; Horowitz 1999). The creation of a uniform model of a pupil or of a citizen was meant to express the aspiration for equality that appears as a basic principle in Israel's Declaration of Independence. This aspiration penetrates the entire history of the educational system in Israel.

However, contrary to expectations, there were still academic gaps among the various ethnic groups. The gaps might be explained by the "fragile state theory" (Fuller 1991), which claims that many developing societies are under pressure to achieve modernism, trying to solve social and economic problems without harming the advantages of the elites. Schools are organized in ways that serve the political and symbolic goals of the state, yet do not necessarily provide *mass* education. The contradiction about the fragility of the state is that education cannot become universal without undermining the status interests of the political and social elites. As a result, development might bring about the broadening of the educational system, yet reform will be superficial for those who need it most.

Mickelson, Nkomo, and Smith (2001) claim that the Ashkenazim (Jews of European origin) in Israel created and preserved their dominance by using political resources, economic power, and cultural capital. According to them, one of the official goals in Israel is to bridge the gaps between the groups that have an advantage and those who lack it, breaking the tight link between ethnic origin and socioeconomic status. In view of these gaps, it was decided during the 1970s to conduct a reform in the educational system. Middle schools were established with the purpose of combining students from various areas of residence and different socioeconomic backgrounds into heterogeneous classes. The hope was that studying together would lead to an equality in academic achievements, or at least to a narrowing of the gaps among the students.

In addition, during recent years a pluralist policy has begun to seep into the educational system. The policy is characterized by more tolerance toward value systems that are not representative of the official policy (Eisikovits 1997; Horowitz 1999). This policy corresponds to the trend in Anglo-American states, whose schools tend to adapt themselves to immigrants by adjusting their rules and their curricula to the needs of the immigrants (Davies and Guppy 1997).

The new openness goes hand in hand with the enhancement of parents' involvement in the educational system in Israel. It serves as an expression of the moderate pluralist policy. This policy enables various groups in the population to present a wider variety of initiatives that express their needs and preferences. Openness to these initiatives brought about the development of special schools or unique classes in regular schools. Within these frameworks, more time is allocated to certain aspects of cultural heritage over other aspects in the socialization process. The instruction strategies are meant to reinforce and maximize studies in these subjects (Hoover-Dempsey 1992; Rawid 1985).

The second aspiration of the educational establishment is to develop excellence, maximizing the expression of students' ability and the acquisition of scientific and technological knowledge. In Israel, as a small country without natural resources, there is a special significance to bringing the limited human resource to its full potential and expanding the portion of scientists and high-tech professionals. This aspiration cannot be detached from the economic globalization that focuses educational reform on the takeover of global market forces. The spreading net of market relations necessitates standardization of information systems in the industrialized countries. Countries act in order to improve their relative advantage through the educational systems (Davies and Guppy 1997; Reich 1992).

Drucker (1993) claims that the old means of production—land, work, and capital—will be overshadowed by knowledge, the key resource of the next century. The new sources for productivity are based on the applications of science and technology for production and distribution. In Israel, the dominance of the socialist ideology has been overshadowed by individualism, personal achievement, and professional direction. Working the land, the means of production that had been a noble symbol of the relation of the people to their country, was replaced because of economic globalization by the employment of Israelis throughout the world in high-tech industries and in professions where knowledge is the capital.

The efforts to integrate aspirations for both equality and excellence and to find educational attitudes and teaching methods that combine them are very complicated. Equality might even come at the expense of excellence or vice versa. Unfortunately, the academic gaps between students from different ethnic and socioeconomic groups persisted. In addition, there was a decline in academic achievement in mathematics and science by Israeli students, who had usually scored high in comparison with other countries. The results of the Third International Mathematics and Science Survey (TIMSS-R 1999) did nothing but confirm the Israeli public's dissatisfaction with the level and quality of education in Israel. This dissatisfaction was expressed in interviews that I conducted with the families of those students who chose to study in the Top science classes. Families indicated their need for a high-level scientific education in order to be able to professionally compete both in Israel and in the world.

As we have seen, the establishment of Top's science classes was made possible by the change in the social and educational policy in Israel from assimilation as an equalizer of opportunities to a moderate pluralist policy (Eisikovits 1997; Horowitz 1999). This policy permitted various groups in the population to present a wider variety of initiatives that expressed their needs and preferences in special schools or unique classes in regular schools (Rawid 1985; Hoover-Dempsey 1992). In the educational atmosphere that was created, the need to answer the parents' yearning for change and for emphasis on scientific education brought about cooperation with the educational initiative of the Top organization.

Figure 10.1 depicts conditions for borrowing reform within the Israeli educational system and among parents and students. However, so far we have discussed only the issues on the left side of the figure. We have yet to explain how the immigrants have become the agents of change and exactly what change they presented.

THE IMMIGRATION WAVE
FROM THE FORMER USSR IN THE 1990s

The right-hand side of Figure 10.1 illustrates three characteristics in this wave of immigration that, in interaction with Israeli society, paved the way for the foundation of the Top organization and its development into a comprehensive structure:

(1) High education level. About half of the immigrants who arrived during the 1990s had advanced educational degrees. This immigration can be seen as a "brain drain" from the former USSR to the Western world. This wave of immigration included Jews from the cities that constituted the centers of culture, government, and science of the USSR. Jews from these regions had preferred immigrating to the United States during the 1970s and the 1980s (Horowitz and Leshem 1998; Sharlin and Elshanskaya 1994).

(2) The scope of immigration. The huge scope of the immigration has given the immigrants power and in fact changed the balance of power in Israeli society. Over a decade, the Jewish population in Israel has increased by 25 percent while the total population in the country has increased by 20 percent. One of the factors that gave this wave of immigration power was its electoral capacity, especially in view of the changes that occurred in the electoral system in Israel. This power is enhanced due to the immigrants' being a "multidimensional group that is rich in human capital" (Horowitz and Leshem 1998) and to their cohesiveness (Isralowitz and Abu Saad 1992; Mesch and Shaginian 1998).

(3) Roots in Israeli society. Another factor that enhanced the power of this group of immigrants was the existence of a substantial group of veteran immigrants who had arrived in prior waves of immigration from the former USSR. According to Gibson (1995), this constitutes a meaningful base for integration into the educational system. Yet, it should be noted that Israeli

Figure 10.1   Immigrants as Initiators of Educational Change

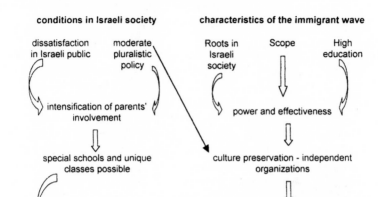

society is characterized by multiple voices. Educated citizens of Western and European descent saw the new immigration in a more positive light than those of Eastern or North African descent who had a lower educational level. The opposition of Arabic society to this wave of immigration stood out (Al-Haj 1993; Isralowitz and Abu Saad 1992).

The combination of these characteristics with the readiness of the local society enhanced the most recent group of Soviet immigrants' influence on the borrowing and transforming process. Contrary to the large immigration that arrived in Israel from countries other than the USSR during the 1950s, which was pushed to the fringes by the policy of assimilation, the absorption conditions of these recent immigrants were different. The change in ideological approach to a moderate pluralistic perception enabled the immigrants from the USSR to preserve their culture in the framework of varied cultural activities conducted in Russian, such as newspapers, literature, clubs, theatre, and independent organizations. The policy of "direct absorption" (decreasing the establishment's involvement) laid the basis for the creation of independent organizations for the fulfillment of the immigrants' needs, financed by the state. This was also the basis for the foundation of the Top organization.

MOTIVE FOR THE INITIATIVE:
THE ENCOUNTER OF IMMIGRANT FAMILIES
WITH ISRAELI EDUCATION

The most meaningful motive for their immigration, noted by the immigrants during their interviews with me, was to ensure a better future for their children (see also Eisikovits 1995b; Sharlin and Elshanskaya 1994). According to Ogbu (1991), the improvement of conditions relative to those in the country of origin constitutes a motive for immigration. For immigrants from the former USSR, this is directly related to education. Acquisition of an education was highly important in Soviet society. These immigrants, who are a product of Soviet education (at least the parents), see a high correlation between level of education and eventual profession. A good profession assures, in their eyes, prestige and better living conditions (King 1979; Zajda 1982). Soviet Jews were characterized by a high motivation for learning in order to overcome the discrimination against them (Eisikovits 1995b).

During the interviews, the immigrants told me they had heard that education in Israel was wonderful. This rumor was further strengthened by a logical deduction: If education was of such a high value for Russian Jews and generally, according to them, they were the excellent students in their classes and winners of Soviet academic Olympic competitions, then in Israel, the "Jewish State," education must be on a very high level.

Immigration for the sake of improving the social and professional future of their children through good education can be seen as part of the process of economic globalization. It should be noted that the sense of commitment to study and to excel for the achievement of social status still exists, though the immigrants are no longer a Jewish minority in Soviet society, but rather new immigrants in Israeli society.

Immigrants explained during the interviews: "We are new immigrants. There is no one to arrange a good job for us. If one wants a good job today in Israel, it is high-tech, and for this one requires an education and good grades."

These background factors influenced the educational integration of the immigrants. They had and have high and specific expectations regarding schooling and its function. Their satisfaction or their criticism derives from their expectations. There were big gaps in their perception of the education they desired and the existing situation in Israel. These gaps were especially big with respect to the sciences and mathematics, subjects that occupied an important place in the Soviet educational system (Eisikovits 1995b) and that also formed for them a meaningful part of their perception of the future in Israel. Students, parents, and even politicians who emigrated from the for-

Figure 10.2   The Encounter of Soviet Immigrant Familes with Israeli Education

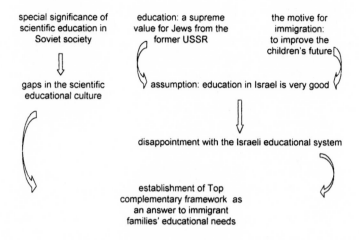

mer USSR indicated the importance of these subjects as part of their cultural background and as an essential basis for success in professional competition and job marketing.

The gaps in the educational culture and the great disappointment with the Israeli educational system created the need for the establishment of a complementary or alternative structure (Figure 10.2).

Gaps in the Culture of Education

The principles that led the Top organization in its work in science classes related to gaps in the culture of education in general and in science education in particular. The organization insists on expert teachers only (that is, teachers with at least a Master of Science in the subject taught). They believe that an expert teacher has the skill to simplify the material for the younger and weaker students on the one hand, and to present challenges to the curious and able students on the other. The expert can develop in them independent and creative thinking and learning through exploration. Such a teacher is not bound by the textbooks. In Israel, however, teachers at the elementary and junior high levels are usually generalists who teach several subjects.

The Top organization believes that it is very important that the same teachers continue teaching their students in every educational stage—elementary, junior high, and/or high school. This way the teachers' responsibility for their students' achievements is clear and they can't blame former teachers. In Israel, there is normally a more frequent change in teachers.

The Top organization also believes in a different distribution of study material. More weight should be placed on the elementary and junior high stages. In what they call "spiral teaching," they gradually deepen the students' understanding according to the latter's age and the dynamics in their class.

In addition, Top sees mathematics as a science and as a basic language used in the study of sciences rather than as a technical skill. Mathematics and physics are perceived as tools for life, teaching students how to think in a logical, organized manner and to seek solutions.

IMMIGRANT TEACHERS:
AN UNUSED HUMAN RESOURCE

The gaps in the culture of education constituted a major factor in the difficulty that immigrant teachers faced in integrating into Israeli schools. One of the reasons for the development of the educational initiative and the foundation of the Top organization was the existence of a big reservoir of scientists and expert teachers who wished to teach and were unable to do so. This reservoir existed for several reasons (Figure 10.3). First, as I learned from discussions in the Knesset, the scope of the Soviet Jewish immigration created a situation where many could not integrate into the educational system due to lack of positions. In addition, cultural gaps and failures in evaluating human resources caused a situation where many professionals were left without work in their profession. A law stating that a principal who employed a Soviet immigrant was entitled to additional funding from the Ministry of Education, which was supposed to help immigrant teachers integrate, missed its purpose. The purpose of the law was to enable the immigrants to acclimatize themselves to the system and then remain in it in existing positions. Unfortunately, there were principals who used the additional hours they received for these teachers while they were entitled to them, and then stopped the teachers' employment and hired other immigrant teachers, which entitled the school to additional hours.

The teachers' organizations strengthened the Ministry of Education's policy that the positions of general teachers who were already part of the system had to be maintained. In this way, they prevented the use of the human capital of the expert teachers who had arrived in this wave of immigration.

According to the interviews I conducted, the system that determined who was entitled to be granted qualification and to teach in Israel failed. The

Figure 10.3   The Encounter of Soviet Immigrant Familes with Israeli Education

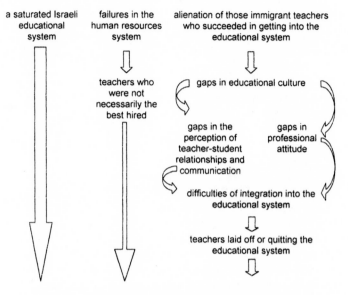

a big reservoir of scientists and expert teachers who establish top as
an answer to their desire to teach according to their culture

committee in charge of confirming the applicants' degrees was not aware of the differences among the institutions of higher education in the former USSR and was unable to discern among the various educational tracks. The committee failed to examine the applicants' work experience, although the places where they had been employed served as evidence of their abilities. The committee recognized the validity of degrees according to the criterion of the number of years of study, without taking into account the fact that fewer years of study might indicate excellence and an accelerated curriculum. As a result of this faulty selection system, teachers whose qualifications were not the best were accepted into the educational system, a fact that caused damage to the reputation of the immigrant teachers.

Finally, cultural differences in educational approaches as well as in the perception of teacher-student relationships and teacher-student communication caused difficulties of integration even for excellent teachers. The teachers encountered many disciplinary problems and they could not implement

Figure 10.4   Penetration of Top into the Formal Education System

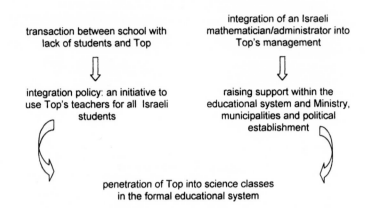

their distinctive educational approaches, as they were immediately told, "In Israel things are not done this way." Being new to the system and strangers to it, they did not dare to insist on their ideas. These cultural gaps caused teachers to leave the educational system. My study reinforces Anderson-Levitt's tentative conclusions (1989, 1990) about the meaningful role of national culture in the conceptions and values of teachers and, as a consequence, in their instructional culture.

FOUNDING OF THE TOP ORGANIZATION AS A COMPLEMENTARY EDUCATIONAL STRUCTURE

The immigrant families' dissatisfaction with the educational system and the reservoir of expert teachers who could and would teach according to their educational culture resulted in an encounter of mutual interests and a search for a solution to their needs. The solution for the immigrants' educational and cultural needs that the conditions in Israeli society made possible was the establishment of a complementary educational structure.

The initiator of Top was an immigrant from the former USSR who was active in political and public life in Israel. He gathered scientists and teachers who had specialized in the USSR schools for excellence in mathematics and physics. During its first stage, the organization was active in evening schools as a complementary and segregated structure for immigrants from the former USSR. The language of instruction was Russian. Studies in these evening schools focused on mathematics and sciences. Subjects related to the preservation of Russian culture, such as chess and Russian literature and language, were also offered.

Thus the dissatisfaction of immigrant families and teachers with the educational system, combined with the power of this immigrant wave and Israel's moderate pluralistic policy, brought about the foundation of an independent organization whose aim was to teach and learn according to the immigrants' culture in a complementary after-school structure.

However, the initiator of Top had a hidden aspiration and a long-term goal, namely, introduction of the culture of scientific education into Israeli society. He hoped to follow the example of the introduction of symphonic music by German immigrants who had established a symphonic orchestra that served their needs and whose influence seeped into and spread throughout Israeli society.

PENETRATION OF TOP INTO THE FORMAL EDUCATIONAL SYSTEM

Ginsburg, Cooper, Raghu, and Zegarra (1990) refer to "institutionalist theory" as an equilibrium paradigm. An equilibrium paradigm assumes that an educational reform or change takes hold as part of a homeostatic system's reaction to a functional lack of suitability that might appear from time to time. It is a natural evolutionary development or adaptation required by a lack of balance in the system or by social needs. This paradigm contrasts with the conflict paradigm, according to which educational reform occurs through conflict and competition among social, ethnic, national, religious, class, and gender groups, whose interests are not compatible. In my opinion, the introduction of Top's educational initiative into the Israeli educational system corresponded to the equilibrium paradigm, as it was a process of seepage and spreading and not a reform that stemmed from conflicts among the various classes.

The process began through relationships that were formed between scientists and teachers who belonged to the Top organization and the principal of a school in the center of the country that was about to be closed due to lack of students (Figure 10.4). In return for increasing the number of students in this school, the immigrants received the opportunity to teach in the formal educational system, according to their own educational culture, during the morning hours. The school soon became a magnet for immigrant students from all over the country. Students who lived far away from the school had to be satisfied with studying during the evenings only. In order to enable them to study in the morning, within the formal educational system, more schools were required.

An Israeli mathematician who was also an educational administrator began to work in Top through the mediation of the Ministry of Absorption. Her job was to help Top cope with the Israeli bureaucracy for the purpose of increasing their activities. Her recognition of their work and her wish to introduce their approach to scientific education to all Israeli students through science classes were aligned with the hidden aspiration of the organization's initiator. Teachers in the Top organization, who saw themselves as carriers of a message and who sometimes encountered difficulties with the Israeli educational system and felt frustrated due to the lack of recognition of their ways, were glad to contribute and prove themselves.

At this point, the organization changed from a separatist organization into a structure that combined veterans and new immigrants. The evening schools began to teach in Hebrew as well as Russian and to accept Israeli students who had been exposed to the organization through their immigrant friends and were impressed by it. Israeli teachers who had Top's students in their classes witnessed the gap in the pace of learning and the scope and depth of understanding between these students and the other students in the class. School principals began turning to Top evening school principals, asking for help when they were short of teachers. This dawning recognition of Top's teachers' work coincided with three other important factors: the growing involvement of Israeli parents, who saw the need to improve the system through educational initiatives; the immigrants' need to establish additional morning structures for their children; and the actions of the Israeli representative whose job was to assist in this mission. Looking at the four factors together, we can understand the basis of Top's penetration into the formal educational system.

Schools in the process of change saw in Top an impetus. Top formed relationships with other schools because of "an encounter of interests," as one of the principals defined it. A relationship was created with schools and municipalities that saw in the organization a means for enhancing their own attractiveness, improving their prestige, and increasing the registration of "potential clients" in their schools attracted by the possible alternative of science classes taught by expert teachers. These schools opened science classes in which the teachers, the teaching attitude, and the lesson contents were determined by Top in collaboration with the school staff. In addition, the integration of Top, which functioned as a brand name for immigrant students from the former USSR, served as a means for enriching the students' reservoir, both numerically and from the point of view of scholastic motivation in general and toward the sciences in particular. As noted by Steiner-Khamsi and Quist (2000), educational reform or change is based on the prestige of the source of the loan, which interests potential clients in borrowing. The prestige of the source provides the legitimacy for the change.

During the first stages of its introduction into the formal educational system, Top had to cope with a lawsuit brought by the teachers' organization and with opposition from various factions in the Ministry of Education and the schools. The opposition derived from the threat that Top's mere existence constituted, since its foundation and activity symbolized potential criticism of the educational system. In addition, there was a conflict, in Top's initial stages, between the educational aspiration toward excellence on the one hand and equality on the other. However, other sectors in the establishment, such as municipalities, educational organizations, factions in the Ministry of Education and even factions in the political establishment assisted Top's activities. This cooperation followed mediation by the Israeli board member, who later became general manager of Top.

Today, in view of the organization's success, there is cooperation between the organization and prestigious schools that recognize its contribution. The number of schools and classes that participate in Top's activities rises in a geometric progression. The distribution among the various age groups has increased and includes middle schools (where the activities had begun), high schools, elementary schools, and even kindergartens.

The Key to Integration

According to my fieldwork, the key to the successful integration of Top teachers into the formal educational system was a set of three factors:

(1) Selection. The teachers that were sent by Top to teach in science classes were chosen very carefully by representatives of the organization, who could evaluate their qualifications and their professional experience. In this manner, Top introduced into the educational system teachers of the highest professional stamp, in contrast to some of the teachers selected through the official Israeli selection process, which adhered to irrelevant criteria and did not know how to evaluate immigrant teachers. The first representatives who entered the system were teachers and scientists familiar with the approach of the schools for excellence in mathematics and physics in the former USSR. Thus an elite unit entered the educational system. The penetration that became possible through Top enabled the teachers to make changes and to work according to their own educational approaches and their professional beliefs. This was made possible through the two following institutions:

(2) Backup and mediation.   The organization's management served as a mediator in the relationship between Top teachers and the school's management. These relations were enhanced due to the functioning of the Israeli mathematics teacher who served as part of the Top organization's management and later on as its general manager and who had a Master of Science as well as experience in educational management. The immigrant teachers could be told, "You do not understand; in Israel things are not done in this way," placing them in a weak position without the tools to cope.

The Israeli manager, however, could say: "I have experience in Israel, I was a teacher here, I performed management roles, and I say that this way is correct." The expression "she backs us up" was often repeated in the teachers' words. This caused principals and their staff to accept things that were "different" yet related to the work of the organization's teachers, perceiving it as part of their way of working. At points of difficulty, the general manager served as a mediating factor between the two sides. Knowledge and recognition of the Top educational approach, on the one hand, and understanding of the difficult points of the Israeli educational system, on the other hand, helped in the development of adaptations that suited the needs and the principles of the organization's teachers, and the school's management and staff.

(3) Entry into the system as a group.   Another factor that helped Top teachers in their introduction into the system was their entrance as a group. This provided them with mutual support, both emotional and social. Newly arrived immigrant teachers were no longer the lonely strangers among the veteran teachers. They arrived as part of a group. Instead of secluding themselves and feeling rejected, they could find the strength to integrate from a position of power, being part of the staff of the Top teachers.

The clustering of teachers also permitted mutual fertilization in the professional sphere. First, there was a fertilization process among teachers who taught the same subject. In addition, communication across teachers helped to develop integrative learning through a coordination of the material studied among the various science teachers. The ability to show flexibility regarding the material studied and to integrate it with other subjects required expertise and a thorough understanding of the material.

THE HYBRID FORMED FROM THE
BORROWED CULTURE OF SCIENTIFIC EDUCATION

A process of transformation of the model according to local logic and creolization occurred in the borrowed culture of scientific education that took root in the educational system (cf. Anderson-Levitt 2000).

As we saw, the educational system changed. However, pressures from the Israeli system brought about adaptation on the immigrants' side as well. For example, one of the most important changes was in the composition of Top classes. Out of loyalty to the original, at first there were excellence classes that accepted only students with high mathematics skills. This requirement created a conflict between the aspiration for excellence and the aspiration for integration. Indeed, the first classes operated in areas characterized by the students' low socioeconomic level and thus they improved the chances of students from these areas to reach achievement levels similar to those of higher socioeconomic areas. However, the dominant aspiration for equality put pressure on the organization to open its ranks. The organization has changed the criterion for participating in Top classes from ability to motivation. In this way, the organization helps in realizing two aspirations—the aspiration for equality and the increase of the reservoir of those engaged in science and high-tech industries.

Other areas adapted by Top are computers/computerization and attitude toward disabilities.

Elements from the Israeli educational culture combined with elements from the Russian educational culture have created in the science classes a hybrid of educational cultures. The integration of Israeli and veteran immigrant teachers within the Top organization and in science classes and the integration of Top teachers in other classes advance the process of transformation in schools where Top operates. The achievements of Top's science classes and the stormy dynamics they have created in the educational system have begun to bring a change of attitudes and a new spirit that find their expression in the rest of the system.

SUMMARY AND CONCLUSIONS

The phenomenon of Top combines aspects of globalization and of multiculturalism. It can be interpreted as a result of the influence of economic globalization on the educational system, as it was caused mainly by public pressure to develop educational initiatives that would enable students to reach academic and professional achievements. This is not a reform that was initiated and promoted by the establishment. If we analyze the process in terms of institutional theory, then the homeostasis was shaken. Pressure from the public on the establishment led to structural and value changes in the educational system.

Davies and Guppy (1997) claim that although multiculturalism is not normally linked to globalization, it corresponds with institutional globalization in one essential way. The mapping of individual identification turns, more and more, around ethnicity, race, religion, or gender. These characteristics come to mediate the relations between the individual and the national community. The multicultural paradox, which is unique to pluralistic and democratic societies, is that in order to preserve social convergence, they have to adopt cultural change (Willis 1997; Duarte 1998). Multiculturalism becomes a product of globalization forces through the use of the vocabulary of citizens' rights, which serves as a global language and a powerful political resource. This language is easily shaped according to the aspirations of individuals or of groups and facilitates cultural change. The immigrants from the former USSR, with the assistance of mediators from the local culture, made very effective use of the vocabulary of this global language in order to ensure as much as they could that their life in Israel would suit their values and beliefs.

An interesting point in this context is that the schools for excellence in physics and mathematics in the former USSR, the roots of Top, were part of a system established by the Soviets in the 1950s to be able to compete with the scientific achievements of the West, mainly in the space and nuclear fields. It is therefore difficult to isolate in this case multicultural elements from those of world culture.

One can summarize this phenomenon and say that the identity of interests or the unique combination of intersecting global and multiculturalist interests made possible the educational initiative that was developed by the immigrants and that then underwent a process of borrowing and transformation by the educational system and the majority group.

REFERENCES

Al-Haj, M. 1993. Ethnicity and immigration: The case of Soviet immigration to Israel. *Humboldt Journal of Social Relations* 19 (2): 279–305.

Anderson-Levitt, K. M. 2000. What counts as the mixed method of reading instructions in Guinea? Fractures in the global culture of modern schooling. Paper presented at the annual meeting of the American Educational Research Association, New Orleans, April 24. ERIC document ED 442 092.

———. 1990. What's French about French teachers' views of the classroom? Paper presented at the annual meeting of the American Anthropological Association, New Orleans, November 30. ERIC Document 340 706.

———. 1989. Memory and ideals in French classrooms. Paper presented at the annual meeting of the American Anthropological Association. Washington, D.C., November 16. ERIC Document 347 127.

Benavot, A., Y. Cha, D. Kamens, J. W. Meyer, and S. Wong. 1991. Knowledge for the masses: World models and national curricula, 1920–1986. *American Sociological Review* 56 (1): 85–100.

Davies, S., and N. Guppy. 1997. Globalization and educational reforms in Anglo-American democracies. *Comparative Education Review* 41 (4): 435–459.

Drucker, P. 1993. *Post-capitalist society.* New York: Basic.

Duarte, E. M. 1998. Expanding the borders of liberal democracy. *Multicultural Education* 6 (1): 2–13.

Eisikovits, R. A. 1995a. An anthropological action model for training teachers to work with culturally diverse student populations. *Educational Action Research* 3 (3): 263–278.

———. 1995b. I'll tell you what school should do for us: How immigrant youths from the former USSR view their high school experience in Israel. *Youth and Society* 77 (2): 230–255.

———. 1997. The educational experience and performance of immigrant and minority students in Israel. *Anthropology and Education Quarterly* 28 (3): 394–410.

Fuller, B. 1991. *Growing up modern: The Western state builds Third-World schools.* New York: Routledge.

Fuller, B., and R. Rubinson. 1992. *The political construction of education: The state, school, expansion, and economic change.* New York: Praeger.

Gibson, M. A. 1995. Additive acculturation as a strategy of school improvement. In *California's immigrant children: Theory, research and implications for educational policy,* edited by G. R. Rumbaut and A. W. Cornelius. La Jolla, CA: Center for U.S.-Mexican Studies.

Ginsburg, M. B., S. Cooper, R. Raghu, and H. Zegarra. 1990. Focus on educational reform: National and world-system explanations of educational reform. *Comparative Education Review* 34 (4): 474–499.

Hoover-Dempsey, K. 1992. Exploration in parent-school relations. *Journal of Educational Research* 85 (5): 287–294.

Horowitz, T. 1999. Integration or separatism? In *Children of perestroika in Israel,* edited by T. Horowitz. New York: University Press of America.

Horowitz, T., and E. Leshem. 1998. Immigrants from the former USSR in the cultural environment in Israel. In *Portrait of immigration,* edited by M. Sikron, E. Leshem. Y. L. Magnes, Hebrew University, Jerusalem (in Hebrew).

Isralowitz, R. E., and I. Abu Saad. 1992. Soviet immigration: Ethnic conflicts and social cohesion in Israel. *International Journal of Group Tension* 22 (2): 119–138.

King, J. E. 1979. *Other schools and ours.* London: Holt, Rinehart and Winston.

Mesch, G.S., and M. Shaginian. 1998. Social integration in a multicultural city: Immigrants from the former Soviet Union in Haifa. Paper presented at the third International Conference of Metropolis Network. December, Zichron Yaakov, Israel.

Meyer, J. W. 1980. The world polity and the authority of nation-state. In *Studies of the modern world-system,* edited by A. Bergesen. New York: Academic Press.

Mickelson, R.A., M. Nkomo, and S. S. Smith. 2001. Focus on gender in comparative education: Education, ethnicity, gender, and social transformation in Israel and South Africa. *Comparative Education Review* 45 (1): 1–35.

Ogbu, J. U. 1991. Immigrant and involuntary minorities in comparative perspective. In *Minority status and schooling: A comparative study of immigrant and involuntary minorities*, edited by M. A. Gibson and J. U. Ogbu. 3–33. New York: Garland Publishing.

Ogbu, J.U., and H. D. Simons. 1998. Voluntary and involuntary minorities: A cultural-ecological theory of school performance with some implications for education. *Anthropology and Education Quarterly* 29 (2): 155–188.

Ramirez, F. O., and M. J. Ventresca. 1992. Building the institution of mass schooling: Isomorphism in the modern world. In *The Political construction of education: The state, school, expansion, and economic change*, edited by B. Fuller and R. Rubinson. New York: Praeger.

Rawid, M. A. 1985. Family choice arrangements in public school: A review of the literature. *Review of Educational Research* 55 (4): 435–467.

Reich, R. 1992. *The work of nations.* New York: Vintage.

Sharlin, S., and I. Elshanskaya. 1994. The impact and outcome of immigration on parenting and child rearing patterns of ex-Soviets in Israel. Paper presented at the 30th International Committee on Family Research (CFR) Seminar of the International Sociological Association. London, April 28–30.

Steiner-Khamsi, G., and H. O. Quist. 2000. The politics of educational borrowing: Reopening the case of Achimota in British Ghana. *Comparative Education Review* 44 (3): 272–299.

Thomas, G., J. Meyer, F. Ramirez, and J. Boli. 1987. *Institutional structure: Constituting state, society and the individual.* Newbury Park, CA: Sage.

Willis, D. B. 1997. An outsider's view inside: 21st century directions for multicultural education. *Multicultural Education* 5 (2): 4–10.

Zajda, J. I. 1980. *Education in the USSR.* Oxford: Pergamon Press.

TOWARD A CULTURAL ANTHROPOLOGY OF THE WORLD?

Francisco O. Ramirez

The traditional domain of anthropological inquiry was human communities uncontaminated by Western influence. A wide range of socialization practices and structures in these communities could not be explained through Western common sense accounts. The discovery of local cultures became an imperative, and the successful anthropological enterprise was an impressive depiction of the myths and rituals of the everyday lives of non-Western peoples. These myths and rituals were central in making everyday life meaningful, and cultural meaningfulness was a core feature of the local orders examined. Cultural meaningfulness was first and foremost a matter of shared recipes or blueprints, a broadly cognitive rather than a narrowly normative matter. All sorts of disputes and conflicts took place, but even these were better understood if one grasped the common frames of reference or interpretation that prevailed in the local community. Local knowledge was the key to understanding local structures and practices, and as the latter varied enormously, so did the content of local knowledge. Local knowledge was at the heart of local culture.

The scope of anthropological inquiry has expanded beyond its traditional domain. All communities, structures, and practices are fair game. This is especially evident in the rise of educational anthropology. Its subject matter is coterminous with educational sociology or for that matter any social science discipline applied to education. Educational anthropologists study schools, students, teachers, pedagogues, curricula, assessments, educational administration, and policy. No one in this volume focuses on the initiation ceremonies that were once the bread and butter of anthropological studies of socialization practices. Indeed, the local emphasized throughout this book turns out to be national more often than not, though nation-states were by no means the units of analysis traditionally favored in anthropology. Lastly, these chapters emphasize the variability and complexity of local educational sense making, but this is now seen in tandem with efforts to cope with translocal educational principles and policies. The latter are commonly packaged as educational reforms.

The focus on sense making constitutes more common ground between the world culture research tradition that I represent and the anthropological perspectives emphasized here. We differ on the locus of sense making, from local communities to nation-states to world society, but we know that people need to act as if "God were not mad." In what follows I first briefly sketch the development of the world culture perspective, emphasizing its unfolding research agenda and its underlying theoretical assumptions. Next, I briefly address some of the specific challenges this book raises for world culture theory, challenges explicitly summarized in the introductory chapter. These include variation rather than homogeneity in educational policies and practices, trends and cycles in educational reforms, and the place of power in world culture analysis.

THE WORLD CULTURE RESEARCH PROGRAM:
A BRIEF EXPOSITION

Twenty-five years ago, a group of sociologists at Stanford sought to make sense of what had been called a "world educational crisis" (Coombs 1968). Primary enrollment growth was the initial phenomenon of interest. Our analyses showed that too much of this growth that took place worldwide could not be accounted for by focusing solely on the characteristics of the nation-states or their educational systems. Primary enrollment growth was found across varying levels of national economic development, types of political regimes, modes of educational governance, religious traditions, and historical legacies. Not only was formal schooling on the rise, but multiple sources affirmed that the value of schooling for individuals and their societies was also on the rise. And it was not just the value of schooling for this sort of individual or for that kind of society, but the value for all individuals and all societies. Despite vast differences in structural characteristics and national legacies, nation-states acted as if they were committed to the educational enterprise. Our research led us to the assumption that nation-states were embedded in a world environment within which expanding education was the just and proper thing to do. To be taken seriously as a nation-state, countries had to expand schooling or at the very least embrace educational expansion as a national goal. To be taken seriously as a national leader, a regime had to identify itself with national educational expansion. Thus, Julius Nyerere would proclaim, "If I leave to others the building of our elementary school system, they [the people] will abandon me as their responsible national leader" (quoted in Thompson 1971).

Our initial studies focused on educational developments after World War II; the phrase "in the post–World War II era" was liberally used in our papers to convey the idea that our findings were conditional on a world environment that itself had changed historically. This recognition led in different research directions. First, there were studies of other educational developments across nation-states to ascertain whether expansion was one of the multiple educational changes that defied societal-level explanations. This research included studies of compulsory school law and educational ministry formation as well as investigations of curricular developments. Second, we studied the late-eighteenth-/early-nineteenth-century origins of mass schooling. Third, we studied nation-state and world society formation itself. The introductory chapter generously covers a lot of this terrain and summarizes many key findings. Some of these findings emphasized common trends, while others continued to challenge the dominant comparative tradition, which imagines societal outcomes to be due to societal characteristics. This dominant tradition included both social modernization and conflict theories, both of which privileged endogenous processes as explanations of development. In various papers we reiterated the simple point that the educational developments of interest were taking place in both more and less modern societies as well as in countries that varied with respect to degrees and kinds of internal conflicts.

The causal motor, we insisted, must be found in the wider world. Nation-states, our studies suggested, were not best conceptualized as bounded entities or closed systems. This is a meta-theoretical premise we shared with a variety of dependency theories and with the world-system perspective of Immanuel Wallerstein. But within the latter perspectives, the less bounded nation-state is seen as affected mostly by power and exchange processes among world actors with their competing interests and interest-driven goals. These perspectives were useful in identifying and interpreting a set of differentiating outcomes among countries. These ideas had the flavor of a zero-sum game: core country developments at the expense of those in the periphery. The kinds of outcomes our studies highlighted—for example, the institutionalization and expansion of mass schooling covering increasingly similar subject matter—were either not addressed at all or were inadequately addressed within these perspectives. Part of the problem was that the actors and their interests and goals were not problematized within the realist tradition embodied in these perspectives. That is, the cultural roots of their identities and goals were ignored. This critique is often made of liberal theories, where the rational self-interested actor is the starting point. Some chapters in this book illustrate this critique. But the critique applies to theories where the rational self-interested class or state is the starting point as well. Even at the level of national societies, it behooves us to focus more clearly on the institutions that give rise to the actors and the interests, instead of assigning the latter too much context-independent entitivity. (See the chapters in Thomas et al. 1987 for material illustrating this point; for a critique of the limited understanding of interests in both liberal and Marxist perspectives, see Bowles and Gintis 1986:20–26.)

Our alternative starting point is to emphasize the degree to which actors, interests, and goals are contingent on the wider world for their identities and purposes. The wider world legitimates some identities, some goals, and some technologies for attaining these goals. Schooling arises as a favored technology for identity affirmation and goal attainment; its intense pursuit by individuals and by states makes sense only in a world that strongly privileges schooling. The worldwide contemporary celebration of education as human right and education as human capital is intelligible only if world educational models are taken for granted. We went further and asserted that the actors themselves—individuals, nation-states, organizations, professionals, and other "modern" experts—owed their validated identities to the wider world culture. That is, the rationalized sense making linked to these entities—the stress on individual empowerment, national development, organizational effectiveness, professional knowledge, and expert advice—is best seen as involving cultural scripts rather than technical processes. The culture at work increasingly fostered both rationalization and universalism, affirming that all individuals and all societies would benefit from schooling. The culture at work, we later asserted, was articulated and transmitted through nation-states, organizations, and experts who themselves embodied the triumph of a schooled world "credential society" (see the chapters in Boli and Thomas 1999).

The research directions we undertook reinforced these premises, which in turn informed further research directions. First, it became clear that educational expansion was but the tip of the iceberg. Many other educational developments spread throughout the world in the last five decades of the twentieth century. There was no viable worldwide movement opposing education for all or demanding the de-schooling of society. For better or for worse, education and society were thoroughly schooled. Second, the earlier debates as to the educability of peasants, the working class, women, people of color, or the colonized had in principle come to an end. Both education as human right and education as human capital doctrines undercut the conservative position regarding the limited educability of most people. Enrollment growth has not been limited to primary and secondary schooling; tertiary enrollment has also expanded throughout much of the world. The result is not a cheerful consensus. On the contrary, new generative tensions arise over both curricula and pedagogy in lower schooling and selection criteria for higher education admissions. As we have repeatedly emphasized, the triumph of egalitarian standards raises the bar and facilitates the discovery of ongoing or new inequalities, all of which can be readily classified as inequities (Ramirez 2001). I shall return to this point later. Lastly, studies of nation-state formation also revealed common trends in some important domains that include the expansion of both the constitutional powers of the state and the rights of citizens, the standardization of nation-state goals, and the increase in the scope of state bureaucracies. Educational isomorphism went hand in hand with nation-state isomorphism. The ideal nation-state committed itself to education for nation building. This world cultural ideal was anchored in world society itself, that is, in the world and regional organizations and associations, which served as carriers and sites of world culture articulation.

To more effectively communicate these ideas, we have asked readers to pretend that they were anthropologists from a larger and more sophisticated planet trying to figure out what makes these earthlings who they are. That is, we ask them to try to ascertain whether there are some shared common sense assumptions or schemes and whether there are some institutional manifestations of these common models of reality. Alien anthropologists are both less likely to reify dominant categories of interest and action and more likely to think of culture along the lines of Geertzian blueprints rather than Parsonian internalized values. If anthropologists from our own planet can now move from the study of small and relatively isolated peoples to dealing with large national entities relatively embedded in a wider world context, why can't an alien anthropologist imagine that the world or planet-earth context has a culture worth analyzing? To borrow from the Spindlers, quoted in the introduction to this volume, are there not worldwide-approved ways of talking to each other about our common interests and our differences? Are there not world myths and rituals worth exploring? Our work suggests that in addition to local and national "local knowledge" there has been a growth in world "local knowledge." This is especially evident as regards schools and schooling.

Another communication strategy has been to ask readers to speculate on what would happen to a newly discovered island, or more precisely to the islanders. We have argued that the application of universalistic standards to this island and its people would produce rationalization along familiar dimensions. There would be political rationalization via the adoption of an appropriate formally democratic constitution, paving the way for entry into the United Nations. Economic rationalization would occur through harder or softer market and development projects and personnel, resulting in all sorts of endeavors for harnessing and upgrading human resources. Social rationalization would call for harder or softer social policies to eradicate inequalities across a broad spectrum. Needless to say, schooling will come to the island and after a period of time the islanders and especially their leaders would learn to emphasize the value of education as a human right and as human capital. We have also argued that had the island been discovered hundreds of years ago, there would instead have been a spirited debate as to whether the natives had souls, and under which imperial jurisdiction it would fall (see Meyer et al. 1997).

We do not argue that the rationalization of our hypothetical island will be solely benign in its effects. There will no doubt be costs as well as benefits. Our point is that this rationalization will involve different assumptions from those that underlined the colonial rush for Africa in the latter part of the nineteenth century. Our island will probably be discussed as a nation-state candidate within the United Nations, rather than as a source of cheap labor and raw materials. The health and welfare of its people are more likely to attract world attention. So too will its familial and religious practices. If these run contrary to world standards about the rights of women and children, for example, the island leadership will face a problem. The same leadership, however, will find allies worldwide when it articulates its vision of its newly discovered national integrity. World models simultaneously affirm the value of both national integrity and human rights. Evidence of this affirmation is found in a range of international treaties, conferences, and organizations. National distinctiveness is permissible, even fostered in some domains, provided world standards not be violated.

These ideas both influence and are influenced by more recent research that seeks to answer the following questions: a) How does the age of a country influence the likelihood of its abiding by world standards? All else being equal, are the newer countries more likely to abide by these standards? That is, are countries born after standards have been institutionalized more likely to sign off on the appropriate treaties or conventions or establish the expected ministries or laws? b) Since countries vary as regards the kinds and degrees of linkages they have with world culture and their organizational carriers, does the level and type of country embeddedness predict higher compliance? c) What forces account for the proliferation of rationalizing and universalistic world standards? Much of the literature emphasizes the dynamics of the world capitalist economy and the interstate system; we have focused on the influence of the authority of education and, more recently, of science.

We have addressed these and related questions in our studies of the rise of the authority of science and its economic and social consequences. We argue that, as with education, the value of science is not a simple function of its utility. Some educational and some scientific developments indeed show positive economic impacts, but parallel developments may actually hinder economic growth (Schofer, Ramirez, and Meyer 2000). Broad faith in education and science is not simply a results-driven outcome that ordinary experience and reasonable people affirm. Nor in our view is this a straightforward case of capitalist power bamboozling the world with science talk masking economic interest. A common pattern in the world we live in is that the problems created by industrial and military science are often answered by the solutions set forth by environmental and medical science. These answers have greater social plausibility than alternative ones in large part because their scientific roots give them much legitimacy. The authority of the science frame cuts in different directions; it can be effectively invoked to attain progress but also to achieve justice, as well as to do both.

To illustrate this point, note that the arguments for the greater inclusion of women in science and engineering education and careers often contend that exclusion is both unjust and inefficient. It is unjust because women have a right to pursue studies and engage in well-paid work without having to endure discrimination. It is inefficient because it involves the underutilization of human resources needed to enhance national productivity and growth. Both these notions are explicitly and repeatedly invoked by the European Commission in making the link between gender equality and science policies in the European Union (ETAN 2000). Not surprisingly, cross-national research on gender-related issues in education increasingly focus on gaps in science or mathematics achievement (Baker et al. 2002) or on gaps in enrollments in science and engineering (Ramirez and Wotipka 2001). Contrary to the widespread sense that these inequalities are routinely reproduced, the empirical evidence shows that the world trend is toward a decrease in both achievement and enrollment gaps between the sexes. But as indicated earlier, the narrowing of some inequalities in some domains makes the egalitarian standard more salient and leads to the identification and discussion of other inequalities. With respect to gender, the rising bar has led to a focus on gender inequalities within scientific careers. But it has also spawned feminist scholarship on the masculine character of science itself and its negative implications for women's voice and identity in scientific careers. Our studies support the thesis that science is a world institution with diverse consequences that are better understood if we pay attention to its taken-for-granted cultural authority (Drori et al. forthcoming).

We also use the world culture framework to make sense of the rise and globalization of the human rights regime (Ramirez et al. 2002). While emphasis on education and nation building in our earlier work dealt primarily with citizenship rights, we now turn to human rights claims and their organizational carriers. From a world culture perspective the questions of interest include a) the identification of trends in human rights treaty ratifications, membership in human rights nongovernmental organizations, and the growth of human rights discourse in general and human rights education in particular; b) the extent to which the age and world embeddedness of a country influences its human rights activities relative to the influence of societal characteristics and historical legacies; and c) the shifting content of human rights from extrapolations of earlier citizenship emphases to new human rights features, for example, from standardized due process to innovative ethno-linguistic rights for minorities.

To summarize, the world culture perspective is a theory-laden research development program. The research has often involved longitudinal and cross-national research designs and quantitative analysis. More recently, though, some of our students have undertaken more qualitatively oriented case studies; see, for example, Berkovitch's (1999) analysis of the changing status of women in the agenda of the International Labor Organization. The underlying theory emphasizes the development of world models of reality and their influence on individual and collective sense making and action. These models are influential to the extent that their scripted identities and goals are enacted through their scripted technologies. These models originated in the West, and the triumph of the West in the twentieth century has led to the intensification of the Western emphasis on both universalism and rationalization. The enactment of these models is especially evident with respect to schooling, a taken-for-granted technology with which to affirm appropriate identities and commitments and to pursue the taken-for-granted goals of progress and justice.

WORLD CULTURE: A BRIEF RESPONSE

The preceding chapters are interesting and important in their own right. Some emphasize the tensions that arise when exported educational pedagogies and curricular emphases clash with local ones, such as traditional Chinese teaching methods, teaching by the book in the Republic of Guinea, or Thai wisdom. Others note that the same educational reforms—parental choice or outcomes-based education, for instance—are reacted to differently across and even within national contexts. And still others emphasize competing models—the World Bank's versus Paulo Freire's, for example. None of the papers pretend to be examining a pristine local culture of schooling, much less local alternatives to schooling. All seek to show the ways in which the local continues to survive or adapt to exogenous influences. This indigenization process itself can command world support, as in European Union support for distinctive locales within its member states or, more broadly, UNESCO support for cultural heritage displays in schools and their communities. Again, it is important to reflect on the forms the native and the local may take. Heritage displays of a racist or sexist or some other character officially designated as offensive are neither world approved nor likely to flourish. As noted earlier, there are limits to distinctiveness.

From the chapters in this book, one can draw the inference that there is much variation in local educational practices and a lot of tension in determining how to enact or resist imported educational policies. From these chapters one cannot draw the inference that there has been an increase in educational variation across countries. This is simply not an issue that the chapters above address, if for no other reason than that the ethnographic materials employed are not designed to gauge long-term historical trends. The introductory chapter notes that the world has indeed grown in population, but technological and communications developments aligned with the universalism and rationalization emphasized in our studies have led to the erosion of geographic and related boundaries. There are not only more schools and more students (in absolute and relative numbers) than there were at the beginning of the twentieth century, but there are also more common ways of envisioning and interpreting the realities of these institutions. The creolization of educational reforms indeed creates differences, but the reforms transcend boundaries and thus create commonalities. There is no creolization of initiation-ceremonies literature because these ceremonies were not subjected to the contemporary universalistic and rationalizing pressures that modern school policies and practices now face.

What do the observed differences emphasized in the chapters above tell us? First, we find more differences at the level of practice than in policy and still fewer differences at the level of principle. This is an expected outcome from a range of perspectives, including the world culture one (Meyer and Ramirez 2000). More importantly, these observed differences suggest that further studies should seek to ascertain whether some dimensions of educational development are more institutionalized and standardized while others remain more contested and variable. Not every American educational fad makes it to the high altar of universal acceptance—not throughout the world and not even throughout the United States. Educational policy and practice more directly connected to broad purposes such as individual welfare and national progress are less likely to be contested than narrower issues such as how to teach a foreign-language program or to manage a classroom. The observed differences among educational systems also suggest the need for further studies that seek to gauge the conditions under which varying aspects of educational development are more likely to be more influenced by local factors than by global ones. Even this nuanced and conditional formulation of the issue may be difficult to investigate, as all sorts of local justifications can be trotted out, even as a world standardized curriculum is enacted. Still, it would be useful to know whether the activation of local educational heroes or traditions is more likely in the humanities than in the sciences, or on the question of teacher autonomy than as regards cost accounting. We have elsewhere directly addressed the issue of educational distinctiveness and the conditions under which educational distinctiveness is more likely to flourish (Ramirez and Meyer 2002). One point worth reiterating is that local educational distinctiveness is more likely when it is part of a broader regional pattern than a truly local one (Cummings 1997). A second point is that distinctive socialization institutions are more viable if they are not also explicitly educational ones. Local schools are more subjected to world standardizing pressures than local kinship or religious structures.

But do not these observed differences indicate different models? Our many references to world models or world blueprints acknowledge that we are not dealing with a singular model free of internal conflict. When our earlier studies identified a European model of the nation-state, we saw that some countries were more likely to emphasize the myth of the individual, while others stressed the myth of the state as the protector of the nation (Ramirez and Boli 1987). But it seems mechanical and misleading to assume that every nation has a distinctive culture of schooling just because no two nations are identical in all school matters. Neither are subnations, communities, and households. But should we assume that any and all behavioral differences imply cultural differences? I think not. But I do think that one can identify competitors, and the most obvious ones share universalistic and rationalizing assumptions less evident in the weaker ones. The case of Paulo Freire is a good example, even though the empirical evidence supporting the pedagogy of the oppressed is not more impressive than that which favors human capital theory. But both perspectives are cast so their analyses of the situation and their proposed courses of action are applicable to all peoples in all societies. Freire offers a pedagogy for the human empowerment of the oppressed. These ideas are presented as applicable everywhere, and that is why supporters and detractors alike can debate the merits of the work everywhere. If the work were cast as exclusively an analysis of the situation of the oppressed in northern Brazil, its cultural distinctiveness would warrant studies of Freire in folklore departments but not in schools of education outside northern Brazil. Likewise, a reading of human capital theory as a culturally distinctive account of school and work in Chicago would have limited its national and worldwide impact. The point is not that these perspectives actually demonstrate new human universals, but that they presuppose them in making their case for human empowerment and human capital formation.

None of this should be especially problematic for anthropologists, provided all of this is understood as world-level cultural production. Models of progress and justice are cultural products; economists and accountants are engaged in culture work as are scientists, educators, and professionals. In many domains, the favored culture work has a universalistic and rationalizing quality. Schooling is one such domain: Note that the educational reforms that travel most extensively have both a universalistic and a rationalizing quality. Some of these reforms rise and fall in their influence, as in the emphasis on mastering specific subject-matter content versus more diffuse goals such as learning to learn. One can study these cycles of influence within the countries where the original school battles were first fought as well as in receptor sites. The papers in this volume suggest that the interaction of the universal with the local will result in distinctive battles. Other reforms have greater staying power, attested to by long-term, enduring trends. The initial reforms that created the schools in the first place have this character. Educational credentialism does not seem to be giving way to alternative criteria for allocating people to jobs. The identification of world trends has been a staple feature of world culture research. But it is just as important to get a handle on cyclical patterns of educational booms and busts. Doing so will further clarify the character of the world culture at work.

Whose culture acquires worldwide standing? Is not the triumph of the West a central dynamic in modern history? Is not imperialism the mother of institutional isomorphism? Is not coercion the most important mechanism through which common educational models diffuse worldwide? Why is world culture theory silent as regards the issue of power? In what follows I respond to these and related questions.

World culture theory has always recognized the Western origins of the world models, as well as the degree to which the triumph of the West has been consequential to its dissemination. There was nothing inevitable about Western ascendancy nor is there any reason to believe that this is a permanent world condition. A world earlier dominated by other powers would surely look different than the one we inhabit today. A different World War II outcome would also have obviously resulted in a different world. World models of progress and justice and their educational implications have clearly been influenced by Western hegemony in general and more recently by the dominance of the Western liberal powers, notably the United States. Thus, some theories have made power/dependency ties the main dynamic to account for many educational developments throughout the world. The power lies in the United States (and earlier with the Western colonial powers) or in international organizations with similar agendas, notably the World Bank. Or, more broadly it is the power of the world capitalist system that accounts for educational developments worldwide. A common underlying assumption is that these developments serve the interests of the dominant actors who shape them, or that these actors block better educational developments inconsistent with their interests.

No doubt one can point to specific instances where this view makes good sense, from the underdevelopment of colonial versus metropolitan schools to educational funding cutbacks in less developed countries due to World Bank structural adjustment policies. But as a general explanation of worldwide common educational trends, this perspective is inadequate. Schools and universities are at least as likely to produce agitated citizens as tranquilized masses. Modern schooling may create a skilled workforce and an infrastructure conducive to foreign investment, but it may also trigger a more rights-conscious set of labor, gender, and environmental social movement activists. As is true with dependency arguments in general, the educational dependency ones fare better when focusing on differentiated outcomes. Common patterns of educational development constitute an anomaly from this perspective. Common patterns imply more commonality than typically serves the interests of the dominant actors. Throughout history the dominant actors did not attempt to control the dominated ones with rhetoric about progress cum development and justice cum equality. On the contrary, yesterday's favored rhetoric explicitly justified a taken-for-granted inequality, an inequality between social classes, between women and men, between racial and ethnic groups, and between peoples or nations. Yesterday's favored rhetoric also discounted the possibility that all nations were capable of development. There were master races, manifest destiny nations, and of course, it was in the order of things that men ruled the roost. Contrast this imagery with today's preferred discourse emphasizing progress and justice for all. It is difficult to explain the sweeping changes in discourse, as well as structure and activity, as mostly reflecting power differences between actors. It is simply not the case that the "lower orders" would have been more effectively controlled through doses of educational co-optation. Schooling has too many amphetamine-like effects to serve as an opiate for the masses.

Instead of assuming extraordinarily incompetent dominant actors, world culture theory assumes that they too are constructed and constrained by the world cultural frame within which they operate. World culture theory focuses not on the power of the actors but on the power of the culture itself. We assume that a culture can be activated throughout a population via different processes and that these include coercion, imitation, and enactment. These processes can lead to nation-state and to educational-institutional isomorphism. World culture theory underemphasizes both coercion and imitation in favor of enactment. There are indeed more powerful actors in the world as well as more successful ones. More powerful actors can more effectively intervene in the affairs of others and, via carrot and stick tactics, redirect their educational system. The rise and demise of vocational education in Africa seems to directly reflect the changing priorities of the World Bank. The educational policies and practices of successful cases are also likely to be imitated, especially when these are imagined to be the means to attaining valued goals, such as development.

So, why do we not emphasize these processes in our work? We have already indicated the limits of interest-driven explanations, and emphases on coercion processes are often linked to these explanations. There are also limits to explanations that emphasize imitation, especially if by imitation we mean constructing or changing the educational policies and practices of one's country so that they look more like those of another country. Take the United States and its educational influence worldwide. Until very recently the newly independent countries of the world did not much embrace the radically decentralized character of American educational (or, for that matter, political) institutions. What was "imitated" was a way of talking about education as a solution to an ever-increasing laundry list of individual and collective problems. This way of talking was heavily influenced by American educational theorists and policy analysts, though their ideas about how to teach or how to assess were often not in place in American schools themselves. We argue that these and other ideas resonate worldwide to the degree that they are in line with world models of progress and justice. To the degree that these ideas appear to be distinctively American, they travel less aptly, although more so than if they were distinct to a less developed country. So, we view the adoption process as involving the enactment of a scripted progress-seeking or justice-caring identity rather than a concrete imitation of American educational features. This in part accounts for the commonly noted loose coupling phenomena—the disconnect between educational goals and means, or a diffuse policy commitment to seemingly incompatible goals. A more thorough imitation would reduce the loose coupling, but also lower the likelihood of the sorts of creolization strategies highlighted in this volume. Institutional isomorphism and loose coupling go hand in hand because we are dealing with the local symbolic enactment of general abstract models. It will not do to dismiss this as "merely" symbolic, given the large number of different indicators of educational growth and change we have identified and analyzed over the decades. But these changes do look more like identity enactment rather than straightforward copying.

Much critical thought in comparative education emphasizes the costs of mindless imitation or the costs associated with coercive pressures toward policies and practices that fail to take local context into account. We share with this tradition of inquiry the sense that what is going on worldwide is not the triumph of optimal educational strategies. We depart from this tradition, however, in postulating the existence and operation of a world culture that strongly influences nation-states and other actors by providing them with legitimated identities they can enact in the pursuit of legitimated goals.

The world culture we postulate is a symbolic order or universe from which much sense making flows. The world culture offers both an opportunity for identity displays and goal postures and a constraint on the forms these may take. The constraints have preoccupied many scholars who fear the homogenization of the world, calling it McDonaldization or nations marching to the beat of the same transnational drummer. This fear, though, presupposes a more tightly wired and internally consistent system than is actually in place. World culture can be activated to justify learning both in English and in a language of one's own, with universalistic education-as-human-capital and universalistic education-as-human-right principles rationalizing varying courses of action. The flexibility is not boundless, though: Favoring English only for boys or a native tongue only for people of a particular class will not fly. Even diverse policies and practices must be rationalized around common models of progress and justice. The world culture is Western in origin, but both its universalistic aspirations and its rationalizing thrust paradoxically facilitate de-Westernization efforts. This paradox poses a challenge for all engaged in comparative education.

REFERENCES

Baker, D., C. Riegle-Crumb, A. Wiseman, G. LeTendre, and F. O. Ramirez. 2002. *Shifting gender effects: Opportunity structures, mass education, and cross-national achievement in mathematics.* MS. under review.

Berkovitch, N. 1999. *From motherhood to citizenship: Women's rights and international organizations.* Baltimore: The Johns Hopkins University.

Boli, J., and G. M. Thomas, eds. 1999. *Constructing world culture: International non-governmental organizations since 1875.* Stanford: Stanford University Press.

Bowles, S., and H. Gintis. 1986. *Democracy and capitalism: Property, community, and the contradictions of modern social thought.* New York: Basic Books.

Coombs, P. 1968. *The world educational crisis.* Oxford: Oxford University Press.

Cummings, W. 1997. Patterns of modern education. In *International handbook of education and development: Preparing schools, students, and nations for the twenty-first century,* edited by W. K. Cummings and N. F. McGinn. New York: Elsevier Science.

Drori, G., J. W. Meyer, and E. Schofer. Forthcoming. *Science and the modern world polity: Institutionalization and globalization.* Stanford: Stanford University Press.

ETAN. European Technology Assessment Network on Women and Science. 2000. *Science policies in the European Union: Promoting excellence through mainstreaming gender equality.* Brussels: European Commission.

Meyer, J. W., J. Boli, G. Thomas, and F. O. Ramirez. 1997. World society and the nation-state. *American Journal of Sociology* 1: 144–181.

Meyer, J. W., and F. O. Ramirez. 2000. The world institutionalization of education. In *Discourse formation in comparative education,* edited by J. Schriewer. Frankfurt am Main: Peter Lang.

Ramirez, F. O. 2001. World society and the political incorporation of women. *Kolner Zeitschrift fur Soziologie und Sozialpsychologie. Sonderheft* (special issue on gender studies, edited by Bettina Heintz) 41: 356–374.

Ramirez, F. O., and J. Boli. 1987. The political construction of mass schooling: European origins and worldwide institutionalization. *Sociology of Education* 60 (1): 2–17.

Ramirez, F. O., and J. W. Meyer. 2002. National curricula: World models and historical legacies. In *Internationalization: Comparing education systems and semantics,* edited by M. Caruso and H. Elmar Tenorth. Frankfurt am Main: Peter Lang.

Ramirez, F. O., J. W. Meyer, C. M. Wotipka, and G. Drori. 2002. *Expansion and impact of the world human rights regime: Longitudinal and cross-national analyses over the twentieth century.* Research proposal.

Ramirez, F. O., and C. M. Wotipka. 2001. Slowly but surely? The global expansion of women's participation in science and engineering fields of study, 1972–1992. *Sociology of Education* 74: 231–251.

Schofer, E., F. O. Ramirez, and J. W. Meyer. 2000. The effects of science on national economic development, 1970–1990. *American Sociological Review* 65: 877–898.

FRANCISCO O. RAMIREZ

Thomas, G., J. W. Meyer, F. O. Ramirez, and J. Boli. 1987. *Institutional structure: Constituting state, society, and the individual.* Beverly Hills, CA: Sage.

Thompson, K. 1971. Universities and the developing world. In *The task of universities in a changing world,* edited by S. Kertesz. South Bend, IN: University of Notre Dame Press.

LIST OF CONTRIBUTORS

KATHRYN ANDERSON-LEVITT is a professor of anthropology and Associate Dean of the College of Arts, Sciences, and Letters at the University of Michigan—Dearborn. She is a former editor of the *Anthropology and Education Quarterly* and has conducted research on cultural knowledge for teaching in France (most recently, in a book called *Teaching Cultures*) and on local and transnational ideals of good reading instruction in Guinea.

LESLEY BARTLETT is an assistant professor at Teachers College, Columbia University. She conducts ethnographic research on popular education, adult education, and the cultural production of race and class difference through schooling.

DIANE BROOK NAPIER is an associate professor in the Department of Social Foundations of Education at the University of Georgia. Her research interests focus on issues of implementation of educational reforms, particularly in sub-Saharan Africa. She has conducted ethnographic and action research in South Africa and surrounding countries since 1990 and she has participated in research and training partnership activities in several South African schools, teacher training colleges, and universities.

BOUBACAR BAYERO DIALLO is an assistant instructor and researcher in sociology at the University of Conakry, Guinea, and a doctoral student at the University of Quebec in Montreal. His research focuses on the academic careers of girls in Africa, particularly of disadvantaged but high-achieving girls. He has also conducted studies of the professional paths of university students in Guinea and has contributed to over a dozen applied research projects in Africa.

THOMAS HATCH is a senior scholar at the Carnegie Foundation for the Advancement of Teaching where he serves as co-director of the Carnegie Academy for the Scholarship of Teaching and Learning and the Carnegie

Knowledge Media Lab. His work includes investigations of the opportunities and barriers to school improvement, research on the development and exchange of teachers' knowledge, and the development of new technologies to document effective classroom practices.

MEREDITH L. HONIG is an assistant professor of education policy and leadership at the University of Maryland, College Park. Her research, teaching, and publications focus on policy design and implementation, decision making in public bureaucracies, organizational learning, urban education, and school-community collaboration.

SUSAN JUNGCK is a professor and chair of the Department of Educational Foundations and Research at National-Louis University in Chicago, Illinois. She has done collaborative, ethnographic, and action research in Thai schools and universities since 1985.

BOONREANG KAJORNSIN is an associate professor and head of the Division of Educational Research and Evaluation at Kasetsart University in Bangkok, Thailand. She has been a project leader in participatory action research and local curriculum development projects for six years.

HUHUA OUYANG is presently head of the English Department at Guangdong University of Foreign Studies, Guangzhou, China, where he has been working as a teacher educator and English teacher for nearly 20 years. His research revolves around preparing in-service teachers for the sociopolitical ramifications of doing curriculum innovation in mainland China. His recent work has been published by journals including *Anthropology and Education Quarterly.*

FRANCISCO O. RAMIREZ is a professor of education and (by courtesy) a professor of sociology at Stanford University. He is also the chair of the Social Sciences, Policy, and Educational Practice Area. His current research interests focus on education, science, and development, on human rights and gender issues, and on universities in comparative perspective. He is co-author of the forthcoming *Science and the Modern World Polity: Institutionalization and Globalization* (Stanford University Press 2003).

DEBORAH REED-DANAHAY is an associate professor of anthropology and Associate Dean of the College of Liberal Arts at the University of Texas at Arlington. She is author of *Education and Identity in Rural France: The Politics of Schooling* and editor of *Auto/ethnography: Rewriting the Self and the Social.* She does research on schooling narratives and is also studying the politics of new educational initiatives of the European Union.

LISA ROSEN is a research associate at the Center for School Improvement at the University of Chicago. Her research and writing addresses the sociocultural analysis of educational policy formation and implementation. Her current work focuses on the cultural dynamics of urban school improvement, particularly the development of professional community in urban schools.

KALANIT SEGAL-LEVIT is a doctoral candidate in the Faculty of Education at the University of Haifa, Israel. She is studying the integration of immigrants from the former Soviet Union into the Israeli educational system in light of the mutual cultural effects of their interaction.

AMY STAMBACH is an assistant professor of educational policy studies and anthropology at the University of Wisconsin–Madison. She is author of *Lessons from Mount Kilimanjaro: Schooling, Community and Gender in East Africa* and is currently conducting research on U.S. faith-based educational initiatives in East Africa.

INDEX